SEA ISLANDS
OF THE SOUTH

by
Diana and Bill Gleasner

Copyright 1980 by Diana and Bill Gleasner
First Printing

All rights reserved. No part of this book may be reproduced without permission from the publisher, except by a reviewer who may quote brief passages in a review; nor may any part of this book be reproduced, stored in a retrieval system or transmitted in any form or by any means, electronic, mechanical, photocopying, recording or other, without permission from the publisher.

Library of Congress Cataloging in Publication Data

Gleasner, Diana C.
   Sea Islands of the South.

   Bibliography: p.
   Includes index.
   1. Outer Banks, N.C.—Description and travel—Guide-books. 2. Sea Islands, S.C.—Description and travel—Guide-books. 3. Sea Islands, Ga.—Description and travel—Guide-books. 4. Natural history—North Carolina—Outer Banks. 5. Natural history—South Carolina—Sea Islands. 6. Natural history—Georgia—Sea Islands. 7. Atlantic coast (United States)—Description and travel—Guide-books. I. Gleasner, Bill, joint author. II. Title.
F262.O96G56   917.5   79-24730
ISBN 0-914788-21-3

All photographs by Bill Gleasner unless otherwise specified.
Drawings and maps by Joel Tevebaugh.
Typography by Raven Type.
Printed in the United States of America by Hunter Publishing Co.

An East Woods Press Book
Fast & McMillan Publishers, Inc.
820 East Boulevard
Charlotte, NC 28203

Warmth of sand
Tug of tide
We, the fragile-winged,
Outsoared the birds
The island . . .
Indelibly ours

For island lovers,
especially those who struggle to
preserve these unique national treasures

Photo by J. Foster Scott

About the Authors

Diana and Bill Gleasner have been a full-time photojournalism team for six of their twenty-one married years. They met at Ohio Wesleyan University, pursued separate careers in teaching and sales and had two children before joining their photography and writing talents on a free-lance basis. A member of American Society of Authors and Journalists, Outdoor Writers of America and the Authors Guild, Ms. Gleasner has written 13 books. Bill is a member of the Society of Magazine photographers and the Outdoor Writers of America whose work has appeared in books, on magazine covers, posters and calendars. The Gleasners have collaborated on hundreds of magazine articles and five Hawaiian guidebooks. They live on a lake in North Carolina where they enjoy waterskiing with their teenagers, Steve and Sue.

Acknowledgments

During the course of our travels we met many memorable people. From the man on the beach who shared the secret of finding sand dollars to Lindsey Vick who sent my bathing suit home, they were helpful in a myriad of thoughtful ways. Islands are fun, but it's the people we met who turned this project into a warmly rewarding experience.

We want to thank those who helped us get started: Paul Phillips of the North Carolina Travel and Tourism Division, Marion Culp of South Carolina's Division of Tourism, Hanna Ledford of the Department of Industry and Trade of Georgia and George Henry of Georgia's Coastal Area Planning and Development Commission.

In North Carolina we are indebted to Natalie Case (who understands rainbows) of Dare County Tourist Bureau, Tony McGowan of the Outer Banks Chamber of Commerce, photographer Foster Scott, pilot Jay Mankedick, Helen Ray of the Carteret County Chamber of Commerce, Mary Lou Pagano (expert alligator sidestepper) of the Greater Wilmington Chamber of Commerce, Patrick Dowling of the Southport-Oak Island Chamber of Commerce, Bald Head Island Corporation, T.R. Vick (who missed a whole night's sleep to turtle watch with us) and Jeri Bartels of the South Brunswick Islands Chamber of Commerce. Miller and Helen Pope of Ocean Isle even threw a party in our honor!

We appreciate the help of the following South Carolinians: Pat McKinney at Isle of Palms Beach and Racquet Club, Ann Harrison of Charleston County Park, Recreation and Tourism Commission, John Ford of the Kiawah Island Company, Suzi Sale of Seabrook Island, Joanne McGill of Fripp Island and Richard Dey of the Hilton Head Island Chamber of Commerce.

It was Georgian hospitality which first sparked our interest in the coastal islands. We're grateful for the cooperation offered by Jenny Stacy and Beth Glass of the Savannah Convention and Visitors Bureau, Anne Pearson of Sapelo Island and Lynn Cheek of Jekyll Island Promotional Association.

Bouquets to Pat and Ed Cheatham who generously offered good advice, Patricia Plyler, marathon typist, and Steve and Sue Gleasner who learned to operate both the washing machine and alarm clock in our absence.

A special thanks to Sally McMillan of The East Woods Press who shared our enthusiasm from the beginning and helped us turn idea into reality.

SEA ISLANDS
OF THE SOUTH

Contents

Introduction—An Island Fantasy	13
Shifting Sands	17
Delights and Dangers	23
The Sea Turtle	29
Island Birds	33
Sea Shells	39
Seafood	45
Islands of North Carolina	49
Islands of South Carolina	93
Islands of Georgia and Florida	127
Glossary	151
Bibliography	153
Index	155
Southern Sea Islands: A Color Portfolio	161

ISLAND FANTASY

Recommending is a hazardous business. Since we've browsed the coast from the Outer Banks of North Carolina to Florida's Amelia Island, we should be able to answer the question, "Which island was best?" We could but we won't, because our "best" will probably not be yours.

Do you prefer desolate beaches to dune-to-dune people? Would you rather peel shrimp on a wooden table or sip she-crab soup by candleglow? Are you interested in old houses, turtles or drawbridges? Do you like to surf, fish or loaf?

The more places we go, the more people we talk with, the greater our realization that the prime ingredient for a successful trip is self-knowledge. Also, it is somebody's law that as soon as we say "X" island has the best shells, the sea life for twenty miles around will pack up and leave. Yet, as leery as we are of making specific recommendations, we would like to share some of the peak experiences we've had researching this book.

Before we launch, you should know what kind of people we are. We were engaged on a windjammer, honeymooned in a canoe and live on a lake. We have an instinctive tropism for islands, prefer natural beauty to neon and love people—a few at a time.

We have taken liberties with these highlights by combining them into an "ideal" day, even though we know these islands weren't meant to be ruthlessly hopped, but to be savored slowly. It bothers us not at all that our breakfast location is a full day's travel from our mid-morning activity. This is true space travel, in the spacey realms of our imagination.

This, then, is our fantasy, our dream choice for a 24-hour period.

—•—

Up before sunrise, we find ourselves out on the beach. All seven feet are firmly embedded in cool sand (a tripod we shall always have with us) waiting for the magic. The place? Ocracoke Island on North Carolina's Outer Banks. A quiet setting for an ethereal experience.

Back to Ocracoke's Island Inn for an oyster omelet. Our taste buds applaud.

Morning is our ambitious time, hours ripe for seeing and doing. First, we learn to fly, dipping and soaring with wind in our hair. Some call it hang-gliding. Slicing the air within view of Orville and Wilbur's Memorial, we call it phenomenal. Jockeys Ridge sand dune provides soft landings for fledglings. Once you have flown on

Sea Islands of the South

your own wing, is anything impossible?

Our mid-morning choice is down to earth, a jeep safari on Kiawah Island in South Carolina, complete with ghost stories, sea shells and an entire convention of birds! We take the kids on this trip. Steve drives one of the jeeps and Sue photographs an alligator. It smiles . . . sort of.

We might as well stop at Kiawah Inn's Jasmine Porch for a cup of she-crab soup to tide us over. Slow-moving fans tempt us to linger, but noon finds us browsing the Hammock Shop (Rte. 17 near Pawleys Island, S.C.) for imported cheese, sausage and crackers. We plan to feast while singing in the wilderness of Hammocks Beach State Park (Bear Island, N.C.). The boat ride (free!) out to the island park gives us a chance to unwind and consider such weighty subjects as whether we'd like to be reincarnated as an egret.

We choose an isolated stretch of sand for our picnic. With our backs to a dune, our souls seaward, we partake. If this were a restaurant table, it would certainly qualify as one with a magnificent view.

After a brief sun-snooze, we'll amble down the beach on Georgia's gorgeous Cumberland Island looking for treasures. We are not fussy. Our shell collection is a haphazard affair, neither organized nor catalogued. But even the cracked ones will later stir us with memories of salt air. Bill does his beachcombing through the lens. However, he always admires my finds and has only two rules. 1. They must fit in the trunk. 2. No living creatures (something to do with one that once expired in a suitcase).

Why is the light so rich, the ocean brilliant with a thousand sparkling eyes? Days on Cumberland Island disappear like sand through open fingers. The waning sun tells us all we need to know about time and tide.

After a shower we would like to breeze-dry our hair on our own private balcony at South Carolina's Isle of Palms Beach and Racquet Club. Here we overlook a seascape in motion. Porpoises are leaping, brown pelicans flying in formation inches above the water. We wonder if they ever collide.

We linger over chilled wine and crabmeat balls at the Emmeline and Hessie Restaurant on Georgia's St. Simons Island. Intense sunset colors are reflected in the water. A fine way to limber up an appetite.

We sink into great wicker chairs at A Restaurant by George in Nags Head on the Outer Banks of North Carolina. We are feeling benevolent having played crossing guard for a duckling in George's parking lot. The pace is leisurely, the food grand. Not inexpensive but, after all, this is a special day.

Our evening choice is live theatre by a traveling troop. At Tidal Hall in Fripp Island we almost feel part of the production of the

Island Fantasy

"Marriage Go Round." Four actors and actresses do everything including change the set. Afterwards we chat with them. They have made us laugh and the play explains everything.

Bushed? Of course we are . . . until we see the turtle. On Ossabaw Island we watch an endangered loggerhead haul her heavy body out of the surf. Seeking her ancestral home, she has arrived to lay her eggs on the same beach where she was hatched. Older than the dinosaurs, she is a page from the beginning of time.

We are told most turtles nest when the moon is full. The Ossabaw moon qualifies. But as it begins to set, we are reminded that all our island-exploring days (and nights) have ended too soon.

Fortunately, our flight of fantasy is a collage constructed of bits and pieces of reality. It's comforting to know that when these memories lose their sharp focus, the islands will be there . . . patiently awaiting rediscovery.

SHIFTING SANDS

Where the Islands Came From And, More Important, Where They're Going

We have the ice age to thank for bringing these sunny coastal islands into being. The polar caps formed during that time continue to melt and, as they do, they add water to the ocean. About 5000 years ago the world-wide sea level, though still rising, slowed to about a foot a century. Some of the old seashore dune ridges, formed when the ocean was lower, were high enough to remain above the water. The land behind them was flooded, either gradually by the rising sea or dramatically during an intense hurricane, and became today's sounds. The isolated strips of sand were stabilized by vegetation which maintained the island's bulk by trapping and holding windblown sand.

These islands are classified as continental islands because they were once connected to the mainland. Those that form a barrier between the sea and the land are called barrier islands. A sea island, in a strict geologic sense, should face the sea, in other words be a true barrier, rather than an inner, island. By tradition Sea Islands is the designation given to a group of islands off the coasts of Georgia and southern South Carolina, (one of which is actually named Sea Island).

We have taken the liberty of gathering all our islands in under the term Sea Islands because of their orientation toward the Atlantic. People may visit museums, play golf or sip mint juleps anywhere; here the draw is the water. It is the ocean that makes the islands special, provides their quintessential appeal.

To us, the term Sea Islands is a real soul-stirrer. It captured our imagination before we ever set foot on the first one, just as it now floods our memories with a wealth of sun-baked images. Finally, if you are looking for hard logic from a writer, consider the case of John Lawson who wrote *A New Voyage to Carolina* in 1709. He threw alligators, rattlesnakes and turtles into a chapter on insects because he admitted "I did not know where else to put them."

These pristine beaches have not been forgotten by time. Endless crashing waves and swaying marsh grasses may seem eternal, but actually these islands are constantly undergoing enormous changes. As soon as an island is formed, it begins to migrate, to alter its shape as well as its vegetation. As long as the sea continues to rise, the islands are expected to retreat in a generally westward direction.

Sea Islands of the South

Sands accumulate in one place this year, another the next. Inlets close, open or simply wander. The Outer Banks (North Carolina's barrier islands) have moved at least 12 miles toward the mainland since the end of the ice age. Change is a far more certain fact of island life than either death or taxes.

Island migration is the term geologists use for what resort owners call beach erosion. We notice that the forces of the universe are awe-inspiring phenomena until they affect people personally in an adverse way. The person whose front yard, which happened to be a beach, has been removed by a passing hurricane does not want to hear about natural laws. Personal property, usually mortgaged, has disappeared. How could it have happened, they wail, as they apply for federal aid.

What is not understood in this case is the nature of barrier islands. They were designed to migrate, just as surely as the Canada Goose. In order to move, the ocean side must edge landward by erosion and the sound side must do the same by growing. It's nothing short of amazing, this ever-changing rearrangement of land. However, there's no point in trying to explain this to anyone whose cottage has just fallen into the ocean.

Shifting Sands

Of Dunes and Marshes

Because the islands are constantly undergoing change, their fragile beauty is a challenge to protect. An understanding of the delicately balanced ecology of an island must precede action. It is folly to build a sea wall every place the waves undercut the dunes. More often than not, this simply speeds the erosion process. In an unending cycle, what happens to the beach vitally affects the marsh, which influences the maritime forests. Understanding these complex interrelationships leads to appreciation. It underscores the mystery of the marsh and hopefully will help us preserve these unique islands.

Dunes begin when windblown sand meets an obstacle such as a single sea oat stalk. A tiny pile of sand forms around it which traps more sand. Gradually the dunelets grow into dunes. In summer the dunes build up and winter storms move the sand back onto offshore bars. By changing shape, the beach can better protect the island from storm damage. Wild waves sometimes pummel the dunes into submission and sweep over them before exhausting their energy. The resulting overwash revitalizes the marsh on the sound side.

The marsh in turn keeps the beach clean by filtering mud from the rivers. Like a gigantic sponge the grass absorbs the force of high tides and violent storms, thereby protecting both beach and mainland from erosion. It is incredible to realize that not too long ago marshes were considered just so much wasted space.

There are many reasons why salt marshes shouldn't be drained and filled to support rows of condominiums. They form the backbone of our coastal environment and are no less than the lifeblood of island ecology. As grasses die, their nutrients feed plankton, fish and shellfish which support larger fish and birds. The material unconsumed in the marsh becomes detritus, a rich organic soup which is carried out to sea on the outgoing tide where it supports fish life. Two-thirds of the commercial fishing catch on the East Coast spends a part of its life cycle in the salt marsh. In fact, the marsh is four times more productive than the most intensely cultivated field of corn.

A tricky interchange of salt and brackish water occurs in this nursery for sea life. Twice a day the tide rolls in, stirring up nutrients and bringing microscopic marine life in to feed. Outgoing water carries nutrients which become part of the food chain. Cleverly designed inlets allow just enough flooding of salt water coming in and the right mixture of waters going out. Think of trying to orchestrate the whole process!

The island was apparently engineered with flexibility in mind. It can bend without breaking. Hurricane-opened inlets relieve stress.

Sea oats and beach grass hold down dunes until man or ocean ravages them.

Shifting Sands

Sand builds up in one area, temporarily abandoning another. Enter man with plans to improve the whole thing by dredging here, building a sea wall there. He thins out the forest to build a hotel not realizing the trees form a canopy over undergrowth which is not salt tolerant. When ocean spray attacks the ground cover, it no longer holds down the soil. This loss of vegetation allows sediment to erode, thereby dooming the few remaining trees. With the trees disappears the habitat for our all-too-scarce island wildlife.

As if we need any more reason for ecology-sensitive development than to know that a marsh turns gold in the sunset and a beach is the best place we know to get a tan.

DELIGHTS AND DANGERS
Climate

Who can fault a climate that offers azalea springs, summer with an ocean breeze, the golden marshes of autumn and January palm trees. While the Outer Banks of North Carolina experience a slightly cooler winter than the islands in the deep South, all benefit from the moderating effect of the ocean. Coastal island winters are warmer and summers are cooler than on the mainland.

North Carolina's Outer Banks, a chain of long barrier islands running from the Virginia line down roughly half the state's coastline, are warmed by the Gulf Stream currents which pass within a few miles of Cape Hatteras. The midsummer mean is approximately 78 degrees Fahrenheit, mid-winter 46. Northerly winds prevail during the cold season, southwesterly in summertime.

The South Brunswick Islands (Holden, Ocean Isle and Sunset) in the southern part of North Carolina average 80 degrees April through October, and 62 from November to March. Summer heat is moderated by the Gulf Stream 50 miles to the east, and none of their distinct seasons are extreme.

Islands in the Charleston, South Carolina, area report mean temperatues of 80 in the summer, 54 during the winter. Perhaps the prettiest seasons are the lengthy falls and springs.

Hilton Head, the southernmost South Carolina island, ranges from the low 80s in summer to the low 50s during the winter. Winter daytime high temperatures average close to 63 degrees December through February. Ocean breezes keep the average maximum temperature below 89 in the summer and above 38 in winter. The island reports 31 days a year when the temperature dips below freezing, 39 days a year when it exceeds 90 degrees. The ocean temperature ranges from 52 in January to 84 in July and August; the ocean swimming season runs from April to October. Hardly a week goes by during the winter season without two or three days of over-60 temperatures.

Jekyll and St. Simons, mid-way down the Georgia coast, have summer daytime temperatures averaging 90 with recent January temperatures averaging in the high 50s. Cold spells are brief and most days the thermometer touches 70. One look at the gardens full of pansies, snapdragons and petunias in January tells the story.

One of the islands' best kept secrets are the gorgeous springs and autumns. The weather is often quite simply perfect, even more

Sea Islands of the South

courtesy of Hilton Head Island

comfortable than summer, and the crowds have disappeared. Island residents from North Carolina to Florida wonder why more people don't take advantage of these ideal conditions. One couple on Hilton Head told us they felt sorry for the tourists who miss what they consider the best seasons this subtropical climate has to offer—winter, fall and spring.

If a Jelly Fish Should Sting

If stung, be sure to fling the jellyfish away from you so it doesn't have a chance to sting again. Jellyfish in the coastal islands don't usually attach themselves to the body but if they do, they're not very difficult to remove.

Standard emergency room treatment for the sting is to wash the wound in hot water, dab it with ammonia and apply a paste of baking soda. But an article in *Salt Water Sportsman* cites meat tenderizer as one of the latest and most effective remedies. Mixed into a paste and applied to sting areas, the tenderizer relieves the pain almost immediately.

Insects

Sea breezes keep insect problems to a minimum, but when the

Delights & Dangers

winds cease, it's a good idea to be prepared. Cutter or Deep Woods Off seem to work best on mosquitoes and Avon's Skin So Soft on gnats. We were impressed with the huge bottle of this product carried by the ranger on Cumberland Island. In fact, everyone we met in the Georgia islands swore by this bath oil. We tried to pin people down as to gnat season and could only come up with "when the wind dies," which could be anytime. When that happens those islanders all smell the same—great!

Hurricanes

Hurricanes are the most terrible storms on earth. Their circular winds blow more than 74 miles per hour with gusts up to 200. They originate over the tropical Atlantic Ocean, in the Caribbean Sea and occasionally in the Gulf of Mexico. Several hundred miles in diameter, hurricanes often move in an erratic path at speeds from ten to 30 miles per hour. These horrendous storms have a unique calm center known as the eye which is unlike any other atmospheric phenomenon.

Hurricanes leave a path of death and destruction in their wakes. People who have not experienced one often believe the winds cause the worst damage. They can certainly be devastating. However, the accompanying floods may be more deadly and the storm surge is most lethal of all.

The debris carried by the winds, everything from pieces of roof to lawn furniture, is worse than the wind itself. People are warned to stay inside during a hurricane to avoid injury from these flying missiles. If the winds die suddenly and the sky clears, take care. You may be in the eye of the storm. When it passes, the winds will be as strong, or perhaps stronger, than before and will come from the opposite direction. Hurricanes often spawn tornadoes which are violent winds moving in a narrow path and causing incredible devastation. Be alert for tornado watches and warnings and be ready to take immediate shelter. Floods brought on by torrential rains can kill quickly. If your area receives a flood warning, leave right away. A few minutes delay may result in tragedy.

As dangerous as they are, winds and floods do not compare with the killing power of the storm surge. This great dome of water crashes across the coastline near the area where the eye of the hurricane makes landfall. The surge, which may be as wide as 50 miles, sweeps everything before it like a gigantic bulldozer. Nine out of ten deaths during hurricanes are caused by this wild wall of water, by far the most dangerous part of the storm.

Hurricanes kill far fewer people now than in days past when they struck with little or no warning. Today we have environmental satellites which detect early formation of the storms. Research planes track hurricanes and a network of National Weather Service

Sea Islands of the South

radars track the turbulent storms. Meteorologists at the National Hurricane Center in Miami, Florida, forecast the storm's course and issue warnings, but these warnings are helpful only if they are heeded. There are plenty of stories about people who ignore warnings, deciding instead to throw a hurricane party and ride out the storm. They are usually about those who are no longer alive.

The hurricane season runs from June to November. A hurricane watch means a storm may threaten within 24 hours. A hurricane warning means it is expected to strike within 24 hours. Learn the history for your area including whether or not your house can be reached by the storm surge. Know where the nearest shelter is and the safest route to it. Keep batteries for flashlights and radio handy. Have emergency equipment available including tools, boards and nonperishable foods.

When your area receives a hurricane warning:
1. Listen for bulletins on your battery-powered radio
2. Fuel your car
3. Watch for rising water
4. Moor your boat securely or evacuate it
5. Protect windows with boards, shutters or tape
6. Tie down outdoor objects or bring them in
7. Save several days' water supply if you plan to stay in your home
8. Move away from low lying areas where the surge will be worst
9. Shut off electrical power at main switch and water at a main valve
10. Use caution evacuating
11. Watch for downed power lines and other hazards

Tides

Tide, the regular rise and fall of ocean waters, is caused by the gravitational pull of the moon. All bodies of water are subject to this pull, but it's only where the land meets the ocean that tides are really noticeable. In the coastal islands tides rise twice a day, the first time because they are directly under the moon and the next time when they are on the opposite side of the earth. When the islands are nearest the moon, its gravity pulls the water away from the earth causing the bulge we call high tide. Twelve hours later the earth has revolved enough so that it is between the islands and the moon. Now the moon flattens the earth creating another ocean bulge or another high tide. The moon moves along its own orbit which causes high tide to occur about a half-hour later each day.

During the full and the new moons, the moon and sun pull along the same line. That causes a higher than usual tide called the spring tide. Small tide changes, called neap tides, happen when the moon

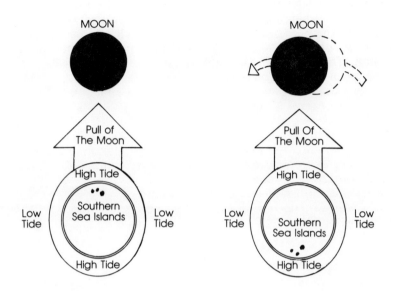

Causes of Ocean Tides

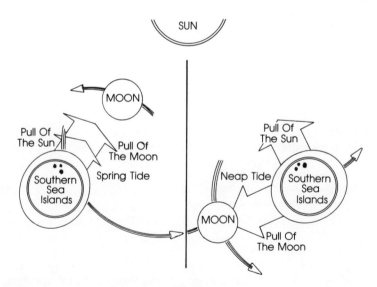

Spring and Neap Tides

Sea Islands of the South

is in its first and third quarters and the moon and sun are pulling at right angles.

The Roman naturalist Pliny wrote of the moon's influence on the tides before the year A.D. 100. But it was only after Sir Isaac Newton discovered the law of gravitation in the 1600s that the physical laws of tides were worked out.

THE SEA TURTLE

She swims in on a wave. Pausing in the surf before moving onto the beach, the turtle holds her head high. A sudden movement or light will send her back to sea. The moon is full; the beach appears safe. Arduously, she drags herself onto the sand. Reaching the edge of the dune, the loggerhead thrashes around to make a shallow area to rest in. Then using hind flippers, she digs a deep hole. Her eggs drop, one by one, into the nest. When she is finished, more than 100 leathery eggs of the size, shape and color of Ping-Pong balls have been deposited. The turtle covers the hole, drops her weight on it several times to pack it down and then scatters the sand in every direction in an attempt to conceal its location. The work is difficult, and every once in a while she rests for a few moments. Gradually her reddish brown shell becomes covered with sand. She turns and lumbers slowly back to the sea, leaving tell-tale tracks that resemble a tractor tread. The turtle lifts her head for one last breath of air and enters the surf a few feet from where she emerged. The whole process takes approximately 45 minutes.

The loggerhead sea turtle has been around longer than the southeastern Sea Islands as we know them. The best guess is that the marine turtle, one of earth's oldest inhabitants, has been in existence for more than 60 million years. That's about twice as long as man. The dinosaurs have come and gone, but the turtle still hauls her unwieldy body up to the dune's edge to lay her eggs.

How has the turtle survived? It hasn't been easy. The species faces an increasing number of predators. Raccoons often follow the distinctive tracks and gorge themselves on fresh eggs from the next. Sand crabs, foxes, hogs and, in some places, people destroy large numbers of eggs. If the nest is not raided, tiny hatchlings emerge about two months later. They face an incredible gauntlet as they dig themselves out of the sand and head for the ocean. Birds swoop down and pick them up, raccoons pounce on them, men gather them for bait. Those who make it to the water face schools of fish and shark that would like nothing better than to lunch on a soft-shelled baby turtle. No wonder only one percent lives to maturity.

Those who survive this highly dangerous time face other hazards. Scientists do not know where the babies spend the first year of their lives. What is known about the adult is that, thanks to streamlined shells, lightweight bones and efficient flippers, they travel hundreds, sometimes thousands of miles at sea. Three hundred pounds is considered large for this species but some have weighed 500,

Sea Islands of the South

Their elongated shells may reach over four feet in length. They live on a diet of crabs, shell fish and jellyfish and have a lifespan of about 25 years.

These turtles spend their entire lives at sea, except for the female's brief excursions ashore to lay her eggs. She does not stay around to see her babies hatch, so everything they do is by instinct. Instinct guides them on vast journeys, leads them back to the same stretch of beach where they were born and teaches them how to make a nest.

Mature turtles have a remarkable ability to survive in spite of serious injury. They have endured natural predators and drastic changes in continents and climate, but civilization has put their future in jeopardy. It is ironic that man has done in just a few centuries what the ravages of millenia couldn't do. He has practically annihilated sea turtles from the earth. Only about 100,000 loggerheads are believed to exist. They are on the U.S. Department of the Interior's endangered species list as well as on the International Union for the Conservation of Nature and Natural Resources list.

When the turtle approaches shore to nest, she is often scared away by moving lights. That means the eggs are dumped at sea where they have no chance of survival. Turtles prefer wide sloping

A Loggerhead sea turtle makes her way back to the ocean after laying her eggs.

The Sea Turtle

beaches so they can easily crawl from the surf to the base of the dunes to nest. Yet careless development has destroyed the dunes. Lights from buildings and cars have frightened turtles ready to lay their eggs and disoriented baby turtles so they lose their way to the sea. (Some recently gave up and settled for a motel swimming pool.) Gradually the wild shorelines of the world are shrinking as human beings take over the habitats that have belonged to the turtles for centuries.

Dead loggerheads are not an uncommon sight on southeastern beaches. Many are victims of shrimp trawlers which drag large nets and leave them on the ocean floor longer than was the case 20 years ago. Turtles accidentally caught in these nets often suffocate because they can't get loose and swim to the surface to breathe.

Most of what scientists have discovered about sea turtles has come from observing and marking turtles when they come ashore to lay their eggs. The tags, which contain an address and sometimes offer a reward for its return, are clipped to the skin of a front flipper. Fishermen who return these tags help scientists learn how far the turtles travel, what their routes are and when they return to nest. They have only theories as to what guides the reptile back to her original birthplace.

What are we, who were given dominion over the fish of the sea, doing to preserve this remarkable species? Fortunately, steps are being taken to set aside areas that will remain wild and undeveloped. Our national seashores and wildlife sanctuaries are an example. Two of the six largest rookeries of the Atlantic States are on Blackbeard and Cumberland Islands, which are protected. Many resorts have selected special lights for beachfront buildings and work to keep moving lights away from the most popular nesting sections of the beach from May to August. Information is being given to vacationers who might disturb the nesting turtles out of curiosity. Some people on Bald Head, Fripp and other islands locate nests and cover them with chickenwire to keep marauding raccoons away until hatching time.

Kiawah Island Company is working with the University of South Carolina on a comprehensive turtle program. Nightly patrols watch for female loggerheads coming ashore to lay eggs. After the mother turtle returns to the sea, the nests are marked and observed. Many eggs are transferred to beachside incubation huts to protect them from predators. Two months later the hatchlings are escorted safely to the ocean. Under these conditions, survival rate is far greater than it would be even in nature undisturbed by man.

The hope is, of course, that these marvelous creatures of the deep will be replenished before it is too late, so they can survive another 60 million years.

ISLAND BIRDS

The Southern islands are on the Atlantic Flyway, a migration route which generally follows the Atlantic Coast. Birds from New England and eastern Canada follow this route to Florida, Cuba and South America. The mystery of how birds know when to start their long journeys and how the instinct which guides them works remains a puzzle to scientists. Because so many are migratory, bird populations on the islands vary greatly in number and species according to the season.

The first group concerned with the protection of birds in this country was a committee of the American Ornithologists Union in 1886. The National Audubon Society, founded in 1905, was an outgrowth of this committee. President Theodore Roosevelt established the first refuge for birds on Florida's Pelican Island in 1903. Today refuges and sanctuaries, scattered along the coast and on the islands, are used as breeding grounds as well as rest stops along the Atlantic Flyway.

Shore Birds

Tern

Terns are members of the gull family although they are more slender and smaller than seagulls. They have red beaks and black heads and sometimes are streaked with red, blue or gray. These agile fishermen dive headlong into the ocean from a height of 30 to 40 feet.

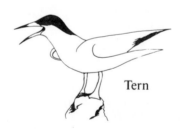
Tern

Plover

Plover

The nervous plover runs swiftly between the wet sand and the high tide line looking for insects or small marine animals. It resembles a sandpiper except its neck and bill are slightly shorter.

Sea Islands of the South

American Oyster Catcher

American Oyster Catcher

The American Oyster Catcher has a black head, a white chest and a red bill well engineered for opening oysters.

Pelican

Pelican

Pelicans seem suicidal but they're just fishing when they dive from heights of 30 or more feet and crash into the sea. They have webbed feet, large pouches under their bills for scooping up fish and a six-foot wingspread. Small flocks fly in long lines sometimes just a few inches above the water.

Black Skimmer

Black Skimmer

Black skimmers have black backs, white undersides and a large red scissor-like bill. Their lower "lip" actually skims the surface of the water as they search for food.

Gull

Gull

Gulls are sturdy scavengers of the sea, not above taking a handout from a fish dock or some bread thrown from a passing ferry. They have webbed feet and a hooked beak. The classiest looking are the laughing gulls with their well-defined black heads.

Island Birds

Sandpiper
The sandpiper is the most common shore bird on the Atlantic Coast. About six inches long, it has greenish legs and light brown back feathers.

Sandpiper

Marsh Birds

Cormorant

Cormorant
A talented underwater swimmer, the cormorant can chase a fish a half mile without surfacing. Afterward, it often perches with wings half-open to dry.

Anhinga
Sometimes known as the snake bird, the anhinga often swims with only head and neck exposed looking very snakelike. This bird has a long tail and a long straight bill. Adults are about three feet long. The anhinga spears its fish, then flips it into the air and gulps it down. Baby anhinga must learn this skill quickly before they become too weak to practice. After a swim, they fan their wings out to dry.

Anhinga

Sea Islands of the South

Ibis

Ibis
The wood ibis is our only American stork. It has a nine-and-a-half-inch oddly curved bill. Ibis fly in V formations with necks and legs extended and with slow, powerful wingbeats.

Heron
The Great Blue is our largest heron with a six-and-a-half-foot wingspread. This solitary hunter walks slowly through the shallows or stands still with head hunched on shoulders waiting for prey to come within striking distance.

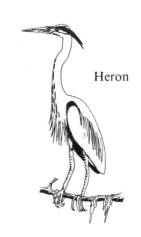
Heron

Osprey

Osprey
The osprey, a fish hawk resembling an eagle but smaller and slimmer, is an endangered species. The osprey must have sharp eyes because it dives from dizzying heights plunging feet first into the ocean. You may see it flying by with a fish securely clamped in sharp talons. The osprey is dark brown with a white underside.

Island Birds

Egret

Egrets are white or mainly white. The American or common egret is larger, about three and a half feet, than the two-foot snowy egret. These birds are protected by various national and international organizations with good reason. They were once driven to the brink of extinction by hunters after their beautiful plumage. Snowy egret feathers brought $32 an ounce on the 1903 New York millinery market.

Egret

SEA SHELLS

The southern Sea Islands are rich hunting grounds for shell gatherers. No wonder the North American Indians used shells for money. They're beautiful! With all our advanced technology we cannot create a seashell. Therein lies the miracle.

Each has its own special design and shape, but all are formed in essentially the same way. A number of specialized glands, using minerals from the food the animal has eaten, form the hard exterior and give it color. Some glands produce the liquid substance that will make the shell; others add a hardening material to make it firm. A different set of glands manufactures the color. If the color is added continuously from one place, a single stripe appears; color from many places results in many stripes. If the color-adding gland operates at intervals, the pattern will be in spots or blotches.

Shelling is a rewarding activity. It's blissfully spontaneous without the interference of rules, requirements or sand fees. Returning empty-handed is not failure; hours spent strolling a beach are a tonic for head and heart. Shell treasures are bonuses. But the search is addictive. Knowing that just around the bend may be the prize find of a lifetime has made many a sheller late to dinner.

The best time to look for shells is winter or early spring after a storm at low tide on a footprint-free beach. No doubt you'll be looking in the summer and may have to do some early rising to find no sign of other humans. Fear not, you shall have shells. They may not be rare or perfect, but they are fine reminders of the magnificence of the sea. Serious shellers are as bad as fishing guides with their "you-should-have-been-here-last-week" stories. The ranger at Cape Lookout National Seashore told us if we'd arrived earlier, we would have seen the beach toe-deep in scotch bonnets. The philosophy of shelling, we've decided, is to use the conditions at hand. Some of life's best moments (and shells) are serendipitous rewards for persisting in the face of less-than-ideal conditions.

Whether you use shells for crafts or simply as souvenirs of your days in the sun, you'll want living animals removed before packing them in a suitcase. Cover the shells with water, bring slowly to a boil and continue boiling for five to ten minutes. Then let the water cool slowly and the insides will fall out. Use a twisted safety pin to extract snails. Very small live shells can be cleaned by soaking them in alcohol for a few days. Wash the shells with soap and water and let them dry in the sun. To soften any encrustations, soak them in a solution of half laundry bleach and half water until you can gently scrape them off. After the shells dry, you might want to rub them

Sea Islands of the South

with mineral oil to bring out the delicate colors and make them shine.

Common Shells Found on Southern Island Beaches

Sand dollar

A live sand dollar is covered by velvety reddish brown spines. According to legend, the five holes represent the five wounds suffered by Christ. On one side is the Easter lily with a center that looks like the five pointed star of Bethlehem. On the other side is the Christmas poinsettia. If you break the shell open, you'll find five white peace doves.

Sand dollar

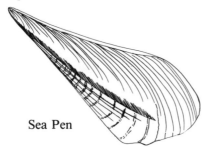

Sea Pen

Sea Pen

The sea pen is an elongated drab brown, very brittle shell measuring usually six to nine inches.

Great Heart cockle

The great heart cockle is pale yellow flecked with rich tan.

Great Heart cockle

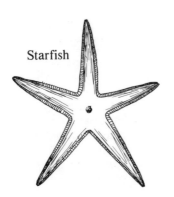

Starfish

Starfish

Starfish, so-called because of their shape, have powerful suction-producing tube feet which are used to open clams and oysters.

Sea Shells

Knobbed whelk

The knobbed whelk is the most common of the conch shells. It has blotches of dark colors on its "shoulders." The whelk egg case, also commonly found, is straw colored and looks like pods on a string.

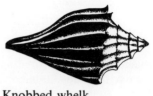
Knobbed whelk

Razor clams

Razor clams

Razor clams are olive green or brown, about two to five inches long.

Angel wings

Angel wings are spectacularly beautiful. The wings of this bivalve, which can measure up to eight inches long, are a chalky white with 26 featherlike ribs on each outer side.

Angel wings

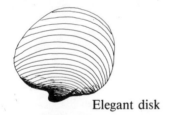
Elegant disk

Elegant disk

The elegant disk is a cream-colored, flattened circular clam, two to three inches in diamter with finely etched concentric grooves.

Lettered olive

A distinctive univalve, the lettered olive has fine brown etchings on a highly polished shell. It measures about two inches long.

Lettered olive

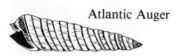
Atlantic Auger

Atlantic Auger

The auger, resembling a fat needle, is pinkish gray, about an inch or two long.

Sea Islands of the South

Oyster
Like snowflakes, no two oysters are ever alike. These bivalves, ranging in colors from dull white to lead gray, are rough, heavily tex-tured and ir-regularly shaped. Sorry, no pearls form in this kind of oyster, but you won't find any better eating.

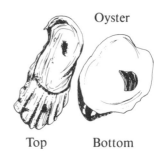

Oyster

Top   Bottom

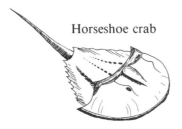

Horseshoe crab

Horseshoe crab
The horseshoe crab is an eight-legged living fossil with a deep reddish brown leathery shell. This animal has remained essentially unchanged for more than 350 million years.

Sunray Venus
The impressive Sunray Venus is a lavender pink clam streaked with dark brown rays. Its sur-face is highly polished and it measures up to five inches long.

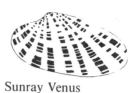

Sunray Venus

Marsh periwinkle

Marsh periwinkle
The small marsh periwinkle is grayish with a spiral line of chestnut dots. It is commonly found on grass in brackish water marshes.

Turkey wing
The turkey wing, a bivalve, has distinct zebra stripe mark-ings.

Turkey wing

Sea Shells

Scotch bonnet

The scotch bonnet is a small univalve about three and a half inches long. Its color is white with spiral rows of brownish orange spots. When North Carolina named this the official state sea shell in memory of early Scotch settlers in 1965, they were the first state to adopt an official state shell.

Scotch bonnet

Callico scallop

Callico scallop

A small bivalve, the callico scallop is white flecked with pink or red.

Coquina

The coquina, a small, rather common bivalve, comes in a variety of colorful patterns with rays of purple, pink, yellow, orange, red, blue and white.

Coquina

SEAFOOD

The Art of Catching, Cooking and Eating a Crab

Catching

Crabbing is great sport. You can enjoy immediate success and proclaim yourself an expert. All you need are a net, bucket, heavy string and some fish heads or chicken necks. Pick your spot, usually a quiet saltwater inlet or creek, and drop in the bait. When you feel a nibble, gently pull up the line and scoop the crab up with the net. That's all there is to it!

Throw back any crab less than five inches wide (from claw to claw) and all females. Females have a rounded apron on the underside while that of the male is narrow and pointed. You can recognize pregnant females easily by the yellowish sponge on their abdomens.

Cooking

Fill in a large pot about two-thirds full of water, add a teaspoon of salt and a half cup of vinegar and bring to a rolling boil. Drop in live crabs (don't cook dead ones) and boil about eight minutes.

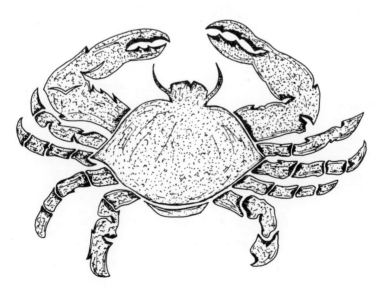

Eating

Eating takes more patience than catching but is worth the effort. Turn the crab over with the underside up. Pry up the apron with a sharp knife and tear it off. Lift off the top shell and pull off the appendages, discarding all but the delicious claws. With a knife, clean out the lungs under the eyes. Scrape off the long whitish, spongelike substance on each side under the top shell and throw away. The body's edible meat is under the semi-transparent membrane. Grip the crab on each side, break in half and pick out the meat. Don't forget to crack open the claws and enjoy them, too!

Nutritional Value of Seafood

	Percent protein	*Percent fat*	*Calories per 100 gms.*
Flounder	18.1	1.4	88
Tunafish	24.7	5.1	168
Clams	11.0	1.7	63
Crab	15.7	2.7	81
Lobster	18.1	1.4	98
Oysters	8.5	1.8	68
Scallop	14.6	0.7	78
Shrimp	18.6	1.6	209

TRANSPORTATION HINTS

North Carolina
The best way to explore the Outer Banks and the rest of the state's offshore islands is by car. Those wishing air transportation will find connecting flights available from major cities into New Bern and Wilmington, N.C. Cars may be rented at the airport.

South Carolina
To reach the islands from Pawleys Island south to Edisto, take public transportation into Charleston. Rental cars and limousines are available at the Charleston airport, and Amtrak has daily scheduled service to the city as well as Greyhound and Continental Trailways buses. Islands south of Edisto including Hilton Head are oriented to Savannah, Georgia. Rental cars can be picked up at the Savannah airport. The city is serviced by train (Amtrak) and bus (Greyhound and Continental Trailways).

Georgia
Those planning to visit the northern Georgia islands via public transportation will arrive at the Savannah Airport, the train (Amtrak) or bus (Greyhound and Continental Trailways) terminals. Rental cars are available at the airport. St. Simons, Sea, Jekyll and Cumberland islands are closest to the Brunswick Municipal Airport which has connecting flights to Atlanta and other major cities. Connections from Atlanta directly into St. Simons can be made via Florida Air. Brunswick is also serviced by Greyhound buses.

Florida
See section on Amelia Island.

Islands of North Carolina

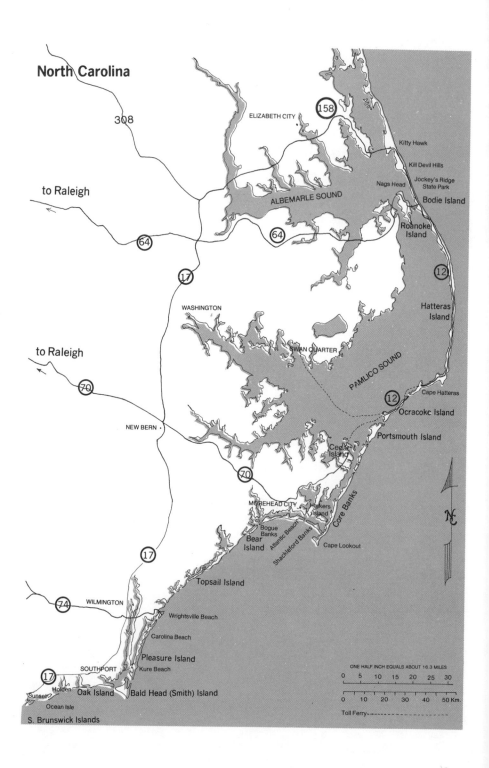

THE OUTER BANKS

How to get there
Car: Rte. 158 from the north. Rte. 64 or 264 from the west. Toll ferry from the south (Cedar Island, N.C. or Swan Quarter, N.C. to Ocracoke Island).
Boat: Intracoastal waterway. Ample docking space, full services.
Plane: Manteo (Dare County) Airport has 3300 feet of paved runways, tie-down facilities, charter plane service, rental cars, sightseeing and flight instruction.
Kill Devil Hills First Flight Airfield. Parking at airstrip's tie-down limited to 24 hours.
Paved strips only on Hatteras Island and Ocracoke Island.

Activities and accommodations
Golf: Courses in Kitty Hawk area.
Tennis: Indoor and outdoor.
Surfing: Excellent conditions at Cape Hatteras lighthouse, site of 1978 U.S. Surfing Championship.
Swimming: Hotel/motel pools and the whole Atlantic Ocean!
Fishing: Excellent! Surf fishing, fishing piers, sound fishing, head boats, charter boats to the Gulf Stream.
Children's activities: Extensive, well-supervised National Park Service program. See Cape Hatteras National Seashore.
Sightseeing: Attractions center on Roanoke Island and Kill Devil Hills.
Camping: Cape Hatteras National Seashore campgrounds and many private campgrounds.
Accommodations: Cottage rentals, hotels, motels, inns, apartments.

For more information
Dare County Tourist Bureau, P.O. Box 399, Manteo, N.C. 27954.
Cape Hatteras National Seashore, Rte. 1, Box 675, Manteo, N.C. 27954.

The Outer Banks are a slim 175-mile string of barrier islands stretching like a protective reef off the North Carolina coast. They extend southward in a crescent from the Virginia line reaching as far out to sea as 30 miles at Cape Hatteras. You can reach them by private plane or boat, but most visitors drive. From the north they come by bridge from Point Harbor (Rte. 158) and Manteo (Rte. 64), from the south by car ferry from Cedar Island or Swan Quarter to Ocracoke Island.

Sea Islands of the South

Would you enjoy the Banks? Yes, if you're happiest in bare feet and a swim suit or jeans. If you like to watch birds or catch fish. If you prefer a quaint inn to a posh hotel. If you want to learn to fly like a bird with the breeze in your face. If lighthouses, gulls and salt air stir your soul.

The purest beach experiences are found at the extreme northern and southern ends of the banks and at stretches along the National Seashore below Nags Head. Most businesses, shops, motels and summer cottages vie for space in the resort developments that cluster around the towns of Kitty Hawk, Kill Devil Hills and Nags Head. You won't want to miss the impressive variety of quality attractions both in this area and on Roanoke Island.

Most visitors head for Kitty Hawk, then travel south the length of the Banks returning to the mainland by ferry from Ocracoke Island. However, it's a shame to miss the area north of Kitty Hawk because the farther off the unpaved path you get, the more likely you are to get to the essence of the Banks. Also, the more likely you are to get stuck in the sand. You won't win any popularity polls by getting in over your hubcaps and asking for help, so stay on the asphalt unless you have four-wheel drive.

Two villages, **Duck** and **Corolla**, lie north of Kitty Hawk, but the real appeal is the natural scene—desolate beach, shore birds, porpoises, ships on the horizon. Follow Duck Road (Rte. 1200 which branches north off Rte. 158 near the Kitty Hawk Fishing Pier) through the cottages of real estate development Southern Shores to Duck, where the road officially ends. This tiny village is not really a tourist destination, but it's a fine place to park and walk the beach. You may come upon some commercial fishermen using four-wheel drive vehicles to winch in their lengthy nets. (We once saw a haul of 10,000 pounds of croakers being taken from the sea.) Or you may see just ocean, sand and sky, which is really what the Banks are all about.

To visit Corolla, you must either walk up the beach 12 miles from where the state road ends, come in your own boat or get permission so an armed guard will admit you through the south gate that blocks what used to be State Road 1152. Corolla consists of a lovely lighthouse, some old cars cultivating rust, a handful of hardy residents, some rambunctious dogs, and a post office where you can hear the latest word on the question of access. The people here are used to the vagaries of the sea but not of bureaucracy. It seems the state never did acquire the right of way for a road, though the public had used it since 1891. A developer declared it private and put up a gate to prove it. That took care of the access from the south. Residents used to heading north to shop in Virginia Beach or Norfolk have been told they can no longer cross Back Bay National Refuge. It's all just a little confusing, but one thing is sure—the issue raises strong feelings! So, if you're looking for some action, get a developer, a few environmentalists and some Corolla residents to gather at the post office one day and see what transpires. Unless you have special permission to get through the gate and a four-wheel drive for the unpaved road, you'd better plan to see Corolla from a sightseeing flight.

Kitty Hawk

Kitty Hawk's name derives, according to one source, from a local Indian reference to the time for hunting geese, "Killy Honker" or "Killy Honk." Whatever the derivation, it is a name made world-famous by the Wright Brothers. Actually their first flight took place four miles south at Kill Devil Hill, which is the site of the Wright Brothers Memorial. But the two towns have overlapped and now they merge into one thriving resort community.

Kitty Hawk was a small isolated village until the north Banks were connected to the mainland in the 1930s by bridges across Currituck Sound. Since then it has grown rapidly. More and more people who escape the cities in the summertime are drawn to the live-and-let-live atmosphere in Kitty Hawk. All kinds of modern conveniences serve a growing population of tourists, and compared to the rest of the Outer Banks, the town is certainly bustling. Kitty Hawk has successfully avoided the boardwalk amusement-type development; the major attractions here are the beach and fishing. In 1969 the North Carolina record striped bass was pulled in from the Kitty Hawk pier.

You may want to stop in at the Dare County Chamber of Commerce Information Center a half-mile east of the Wright Memorial

Sea Islands of the South

Bridge. They can give you a list of realtors to contact about cottage rentals. (9-5 daily. P.O. Box 90-D, Kitty Hawk, N.C. 27949.)

Kill Devil Hills

Why the name? There are various explanations, most of which center on the quality of the imported rum consumed here which at one time was bad enough to, you guessed it, kill the devil. As for the hills, take a gander at the dunes; they actually move due to the windy conditions. Of course it was the wind that attracted the Wright Brothers, and because they came here it's become a must-stop for all visitors. A pilgrimage to the scene of the world's first flight will give young and old a new appreciation of aviation history. Be sure to see all three sights, the **Wright Brothers Memorial,** the **Visitors Center** and the **First Flight Airstrip.** They are within walking distance of each other, two miles south on U.S. Bypass 158.

Wilbur and Orville Wright were in the bicycle business in Dayton, Ohio, but their overwhelming interest was in flight. After gathering information about lift and balance and experimenting with the warping and twisting of wings, they decided to test their theories. They checked the weather bureau in Washington, D.C., for a location with fairly constant winds. All statistics pointed to Kitty Hawk. In 1900 Wilbur Wright wrote, "I am intending to start in a few days for a trip to the coast of North Carolina . . . for the purpose of making some experiments with a flying machine. It is my belief that flight is possible."

Some local residents probably concluded the men were slightly daft as they spent one fall day after another flying their glider as a kite. They kept improving their crafts until they were able to lie prone on the lower wing while airborne. In the process they made

Wright Brothers Memorial, Kill Devil Hills, N.C. (Photo by J. Foster Scott)

North Carolina

more than a 1000 glider flights from the top of Kill Devil Hill setting records for distances and time aloft. By 1903 they were ready to try out a motor-driven machine from the level sand a few hundred feet north of Kill Devil Hill. The age of aviation dawned on December 17, 1903, when their heavier-than-air machine traveled 120 feet staying in the air about 12 seconds. Their momentous experiment was the first successful powered man-carrying flight in history.

Your first stop should be the **National Park Service Visitor Center.** There you'll learn the story of the Wright Brothers and see the full-scale reproductions of the 1902 glider and the 1903 motor-driven plane. Ask for a schedule of talks and kite-flying demonstrations (on the brothers' aviation experiments) which are given periodically. Excellently designed exhibits preserve the rich heritage of this remarkable time, and a good selection of books on the subject is available here. You'll come away with a new understanding of flight and appreciate Orville's awe when he wrote in 1903, "Isn't it astonishing that all these secrets have been preserved for so many years just so we could discover them." (Summer 8-6, Winter 8:30-4:30. Superintendent, Wright Brothers National Memorial, Rte. 1, Box 675, Manteo, N.C. 27954, 704-441-7430.)

A short walk from the visitor center you'll find reproductions of the living quarters and hangar buildings used by the Wrights. They're furnished with tools and other items of the time; taped messages give some of the interesting details of their lives here. Next to these wooden buildings is a field where the first flights took place. A large granite boulder marks the spot where the plane first left the ground, and numbered markers indicate the distance of each of the four flights made that historic day.

A long walk or short drive will take you to the **Wright Brothers National Memorial,** a 60-foot granite pylon atop a 90-foot dune. In 1903 Wilbur said, "These hills are constantly changing in height and slope, according to the direction and force of the prevailing winds." When the decision was made in the late 1920s to establish some sort of memorial to the Wright brothers, it was discovered the dune had moved 50 yards to the southwest though its appearance hadn't changed from 1903. Before they could begin building the shaft, the hill had to be stabilized. Today constant tending of the beach grasses that anchor it in place is required to keep the hill from moving on. Feeling energetic? Climb to the hilltop and collect your reward, a fine view of the Banks, the sound and the sea.

Nags Head

Nags Head, just south of Kill Devil Hills, is a recreation center for those who love the air as well as the sea. Visitors swim in the ocean, roam the beach, fish or rent a boat. Those who are quite

Sea Islands of the South

content to spend their days doing exactly nothing find there's no better place to do it. But for those who want to fly, with or without a plane, Nags Head is the spot.

As to the origin of the town's uneuphonious name, you get three choices. One legend has it that a horse left tied to a tree managed to hang himself and stayed there a bit longer than modern sanitary conditions would dictate. Legend two says the colonists named it for the highest point on the Scilly Isles which was their last view of the mother country as they set sail for the New World. Legend three is most colorful since it comes from pirate times. It seems that not all the pirates were at sea. Some stayed on land, tied a lantern around an old nag's neck and led it slowly up and down the dunes. This created the illusion of a ship riding anchor in a sheltered harbor. A ship at sea seeking just such a haven would head for shore where it would promptly run aground. The land pirates would then take over, thoroughly looting the vessel and doing away with its crew.

Pirate tales abound in this area. A few of these unscrupulous buccaneers made deathbed confessions claiming knowledge of Theodosia, the daughter of Aaron Burr, who disappeared at sea late in 1812. They said they had seized the ship which was headed for New York, taken the valuables, murdered everyone aboard and left it adrift. Years later a portrait of a young woman who looked remarkably like Theodosia was discovered in a Nags Head cottage. The old woman who owned it said it had been salvaged from a deserted schooner that drifted in to shore early in 1813.

Not all Nags Head history is so grisly. In the 19th century the area was a well-loved summer retreat for wealthy mainland planters and their families. Malaria was a warm weather disease, though no one at the time had yet associated it with the mosquito, and they sought the fresh salt air to avoid it. They also had a good time. For the duration of the summer mainlanders joined the hermits and shipwreck survivors who made this isolated strip of sand their permanent home. The area soon became quite a fashionable community with spacious summer homes and even a grand hotel to cater to the seasonal influx.

The great dunes are the most distinctive feature of this narrow spit of land. Their formation is credited, at least in part, to the colonists who destroyed the natural cover of the island by logging and stock grazing. A 1980-acre tract is on the National Registry of Natural Landmarks because it contains the transition from shifting, barren dunes to forested, stable ones. A fine example of a maritime forest which has grown up in the shelter of the old dunes is Nags Head woods. When the dunes were threatened by development, Carolista Fletcher Baum (granddaughter of Inglis Fletcher, the novelist) saved them by placing herself in the path of a bulldozer. She finally convinced the state that the dunes, thought to

North Carolina

be similar to the "marching dunes" of North Africa, were unique and worthy of preservation.

The result of Ms. Baum's efforts to conserve the dunes is **Jockeys Ridge State Park** just north of Nags Head on Rte. 158 bypass. (The parking lot is north of the park.) Some say the dunes once made a natural grandstand for horseracing, hence the name. The Ridge is actually two large dunes, the lower one parallel to the highway, the higher one (approximately 140 feet) in the background. Recently acquired by the state and as yet undeveloped, this 300-acre parcel of land contains the largest natural dunes on the eastern coast.

What do you do on a dune? Kids of all ages know instinctively this is a place for bare feet, for climbing up and sliding down. The ridge is ideal for photographers who want to capture wind-whipped sand patterns and exhilarating views. For some it's the only place to fly, and the only way to go is without an airplane.

Sound risky? Actually, Jockeys Ridge is an extremely safe platform for the modern-day Icaruses who want to soar with the breeze in their ears. A few quick steps and they're off. From the peak of the ridge, the equivalent of a 13-story building, they swoop and glide silently like giant shore birds until they come to a soft landing at the base. Beautiful!

Is it difficult? Not according to John Harris who runs the hang gliding school here, which incidentally is the largest one on the East Coast. Not so incidentally, John was the first person to ever hang glide off Grandfather Mountain in western North Carolina. Whether expert or first-timer, whether your landings are smooth or terrible, you'll find the steady winds off the Atlantic and the soft sand create ideal conditions for improving your skills. More than

Hang gliding over North Carolina's spectacular dunes (Photo by J. Foster Scott)

Sea Islands of the South

2500 students receive instruction here annually and lessons are given year round. Join a beginning class which gets you off the ground and flying, or sign up for a fledgling six lesson package. It won't be long before you're able to use the thermals and ridge lift air currents to sail about 200 feet. Wear long pants and tennis shoes. The school provides helmets and fully qualified instructors, all certified by the U.S. Hang Gliding Association. An added bonus is that most days when the wind is good you'll see Francis M. Rogallo, the man who invented the hang glider or the Rogallo wing, on the ridge. Make your arrangements for lessons or stop in to browse in the unique gift shop packed with all sorts of air-oriented items at Kitty Hawk Kites which is just across from Jockeys Ridge. (P.O. Box 386, Nags Head, N.C. 27959. 704-441-6247 or 441-7575.)

If you prefer your flying in a plane, you can make arrangements for an aerial tour of the Banks at Kitty Hawk Kites. What better place is there to take off than within sight of the spot where aviation history began? A half-hour tour in a Cessna 172 provides a plane's eye view of Jockeys Ridge, the Lost Colony and the long fragile ribbon of sand known as the Outer Banks. Most days the water is clear enough to see sharks, manta rays, porpoises and a good number of shipwrecks resting on the ocean floor. You'll soon understand why the area is called the graveyard of the Atlantic. Longer tours can be arranged which go north to Corolla or south to Ocracoke. (Remember Corolla is difficult, at best, to reach by land.) The Ocracoke flight includes a half-day sailing trip around the island, a champagne lunch and return to Nags Head via the ghost town on Portsmouth Island and a dip of the wings to the wild ponies of Ocracoke. (Mid-May to Labor Day. Kitty Hawk Aero Tours, P.O. Box 386, Nags Head, N.C. 27959.)

Roanoke Island

Twelve-mile-long Roanoke Island is surrounded by four sounds — Albemarle, Croatan, Pamlico and Roanoke. It's reached by U.S. Highways 64 and 264 from the west and U.S. 158 from the east.

The French explorer Giovanni da Verrazano who sighted the Outer Banks in 1524 was convinced that the sounds were the Pacific Ocean or the "oriental sea" as he called it. In 1584 men sent by Sir Walter Raleigh to explore this area returned to England with glowing reports of mild climate and rich soil. Inspired by their enthusiasm Raleigh sent English colonists here in 1585. This journey accounts for the fact that the first child of English parentage was born on Roanoke Island more than 30 years before the Pilgrims landed on Plymouth Rock. The settlers built a fort on the north end of the island, but their problems multiplied. There was sickness, a shortage of supplies and trouble with the Indians. When English supply ships, delayed by war with Spain, finally did get back to the colonists three years later, they found no one. England's first attempt to colonize the New World had failed. The fate of this brave band of settlers remains, to this day, a mystery.

The story of the Lost Colony is told at the **Fort Raleigh National Historic Site** on Rte. 64 three miles north of Manteo. Stop in at the visitor center to examine the exhibits and be sure to catch the film about the English attempt at settlement. Then visit the restored fort and walk the **Thomas Hariot Nature Trail** which starts nearby and winds through the woods to Roanoke Sound and back. Before leaving the National Historic Site, stop in and browse in its shop which has crafts as well as a good selection of books on the area's history. (Fort Raleigh is open mid-June to mid-August from 8:30-8:15, on Sundays 'til six; rest of year 8:30-4:30, closed Sundays. Free. Superintendent, Cape Hatteras National Seashore, Box 457, Manteo, N.C. 27954. 919-473-2111.)

Near the historic site are two very popular visitor attractions, **Lost Colony**—a dramatic performance played in the Waterside Theatre— and **Elizabeth Gardens.** The gardens, on Rte. 64 north of Manteo, are a memorial to the original English colonists who mysteriously disappeared. This unique American garden has a variety of indigenous shrubs and trees, herbs and wildflowers. Many priceless pieces of statuary adorn the site, the Gate House has a collection of antique furniture and there's an ancient live oak believed to have been living when the first colonists arrived in 1585. (Open all year, 9-5. Charge.)

Elizabeth Gardens (Photo by J. Foster Scott)

Sea Islands of the South

The **Lost Colony,** a historical drama, is billed as America's first unsolved mystery. This is the story of the colonists including their loves, heartbreaks, and perseverance against great odds. The thoroughly professional production, written by Pulitzer Prize winning dramatist Paul Green, is produced by the Roanoke Island Historical Association in cooperation with the State of North Carolina and the National Park Service. Since the theatre is open-air, mosquito repellent and sweaters are recommended. The performance starts at 8:30 and is shown nightly from mid-June until the end of August except on Sundays. Unpaid reservations are held at the box office for pickup until 7:30 P.M.(charge). Reservations are suggested and may be made by mail. (The Lost Colony, Box 40, Manteo, N.C. 27954 or call 919-473-3414.)

Roanoke Indian Village, half a mile northwest of Fort Raleigh on Rte. 64, is the scene of demonstrations of Indian techniques for tanning deerskin and chipping arrowheads. With its stake fence, wattle huts and dancing circle, this recreates the type village found by the colonists in 1585. At night songs and dances are taught around the campfire. A campground is located next to the village. (919-473-2463)

The **North Carolina Marine Resources Center** is northwest of Manteo on Rte. 64. Turn left when you reach Rte. 116 and follow the signs to the airport. Displays on underwater archaeology, marine ecosystems, movies on various marine themes and a fascinating schedule of activities from bird walks to fishing classes make this an interesting stop. Here you'll get a good closeup view of some creatures too shy to be seen elsewhere like loggerhead sea turtles, longnosed gar, lobster and a variety of salt and freshwater fish. An exhibit of live sea folk like starfish, sea urchins and sand dollars invites you to touch and hold. (Free. Marine Resources Center, Roanoke Island, P.O. Box 967, Manteo, N.C. 27954, 919-473-3493.)

The towns Manteo and Wanchese were both named for Indian chiefs. Manteo at the north end of the island is the county seat where you'll find the courthouse, medical center, harbor with public boat ramp and the Dare County Tourist Bureau. (P.O. Box 399, Manteo, N.C. 27954. 919-473-2751.) Charter plane service and rental cars are available at the Manteo airport where you can sign up for sightseeing tours or flight instruction. Wanchese on the southern end of Roanoke Island is a commercial fishing community with a public boat ramp and excellent harbor facilities.

North Carolina

CAPE HATTERAS NATIONAL SEASHORE

Cape Hatteras National Seashore extends 70 miles from South Nags Head to Ocracoke Inlet and includes four sections—Bodie Island, Pea Island National Wildlife Refuge, Hatteras and Ocracoke Islands. The islands are connected by bridge or ferry and you can drive the length of the park on Highway 12, a relatively narrow paved road with soft shoulders. This part of the Outer Banks has been saved from the threat of commercial development. Here the true flavor of the Banks is undiluted. Eight small villages, not part of the park system, reflect the culture of the independent souls who originally populated the area, but the seashore remains wild and free. Fishermen are drawn to the superb action offshore, bird watchers come to photograph snow geese and campers look forward to strolling the moonlit beach. Families participate in park programs that include everything from a canoe trip through the marsh to campfire discussions about shipwrecks.

Here, wind, sand and sea contend with each other as they have for eons of time. The profile of this long strip of sand is undergoing change, sometimes subtle change, sometimes dramatic. The ocean rules with a force that is awesome. Man can build beautiful seaworthy ships, the corps of engineers can construct expensive jetties and guidebooks can tell you where the inlets are, but the sea can change it all with one violent storm. This is part of the essential appeal of the Banks and why visitors say they gain perspective on their lives after a stay here. Time and tide won't wait around for you any more than they did for Adam and Eve. Your worries are as important in the grand scheme of things as a speck of sand. So relax, swim, sun and stroll long stretches of lonely beach. Look to the sky. Appreciate!

Bodie Island

Stop at **Whalebone Junction Information Center** (south of Nags Head) for accommodation information and a schedule of activities. This is the first recreational area of its kind in the National Park Service and its offerings are of as high quality as they are diverse. Bird lovers will want to park adjacent to the **Bodie Island marshes** in the lot just beyond the park entrance for a glimpse of egret, heron, glossy ibis and other wading birds.

Coquina Beach, one of the best anywhere, is eight miles south of U.S. 158. Turn left at the sign off N.C. 12 to find picnic shelters, bathhouses and lifeguards on duty from mid-June to Labor Day. At the southern end of the parking lot you can inspect the remains of a real not-so-live shipwreck. Nice to know that all on board sur-

Sea Islands of the South

vived when the *Laura A. Barnes,* once a proud four masted schooner, became stranded on a sand bar in 1921. Her weathered remains pose patiently for photographers.

A must stop for families is the children's activity center at Coquina Beach called the **Sandcastle.** The interpretive and recreational programs (offered every summer day from nine to five) can add many memorable moments to your vacation. Want to learn to catch a crab, create mobiles with shells and driftwood, make a fish printing, surf or snorkel? Care to learn all about sand, beach wildlife, hear shipwreck tales, sing songs of the sea around a campfire? Sign up at the Sandcastle. It's supervised, it's fun and it's free! (919-441-6642)

Bodie Island Lighthouse and Visitor Center is opposite Coquina Beach eight miles south of intersection U.S. 158 and U.S. 64. The rich offering of activities includes early morning bird stalks, canoe trips into Pamlico Sound and evening star watches. Everything is free except the twilight cruise aboard the *Oregon Inlet Queen.* (You can purchase tickets at the marina.) If your schedule doesn't permit indulging in these activities, take a few moments to browse at the natural history exhibits in the center and view the 15-minute slide show on the National Seashore. A short self-guided nature trail through the marsh starts near the lighthouse. The Bodie Island Light, a functional navigational aid, is not open to visitors. Built in 1872, this 163-foot lighthouse is the third in the immediate vicinity. The first was undercut by erosion and the second destroyed by Conference soldiers during the Civil War. (Visitor center is open daily 9-5, 9-6 in the summer.)

Oregon Inlet Fishing Center, just before the bridge and nine miles south of Whalebone Junction on Route 12, is a National Park Service concession. No wonder this is known as the billfish capital of the world. Five to six hundred pound fish are consistently hauled over the transoms of one of the largest fleets of sports fishing boats on the mid-Atlantic coast. In 1974 a 1142-pound Atlantic blue marlin set a new all tackle world record here. If it's fish you're after, you're in the right place. Even the non-fishermen will want to be on the docks between four and six in the afternoon to watch the boats unload their catches.

If you want to try your luck, you may use the public boat launching ramp or choose from half-day or full day inlet-sound trips or offshore fishing trips to the Gulf Stream where the big ones lurk. Reservations and booking arrangements, preferably well in advance, should be made at the booking desk inside the fishing center (919-441-6301) or with the individual captains. (A list is available at the desk.)

The **Oregon Inlet Restaurant** serves a speedy fishermen's breakfast, makes up box lunches and serves evening meals with the emphasis on seafood. (May through October.) When you return

North Carolina

with your boat full, a fish cleaning service picks up the catch, cleans, packages and stores it. For a relatively inexpensive but rewarding fishing experience, book a place on the *Oregon Inlet Queen* for a half day of bottom fishing. (All bait, ice and tackle is provided.) If you'd rather not fish but enjoy cruising, take the twilight cruise aboard the *Queen* which shoves off at 6:30 sharp. *(Oregon Inlet Queen,* c/o Oregon Inlet Fishing Center, P.O. Box 533, Manteo, N.C. 27954. 919-441-6301.) If you just want to throw a line in, there are free fishing catwalks on the south approach to Oregon Inlet Bridge.

Pea Island Wildlife Refuge

Pea Island Wildlife Refuge, managed by the Fish and Wildlife Service, is adjacent to N.C. 12 between Oregon Inlet and the Village of Rodanthe. Pea Island was once separated from Hatteras Island by New Inlet but storms filled the inlet and joined the two stretches of sand. The changing geography makes no difference to the refuge's main attractions—birds. Pea Island, a midpoint on the Atlantic flyway, is a much used feeding and resting area for numerous species of wintering waterfowl. Established in 1938 primarily as a preserve for snow geese, these 5915 acres of freshwater ponds, tidal creeks, bays, low sand ridges, salt marshes, beaches and dunes are a haven for photographers and bird lovers. More than 260 species of birds have been identified as "regulars" with another 50 species having been seen after storms or because of some other unusual happenstance.

Pea Island provides a habitat for the endangered falcon and

Sea Islands of the South

brown pelican and is an important wintering ground for whistling swans, snow geese, Canada geese and many species of ducks. Depending on the season, you may see impressive wading birds—heron, glossy ibises, snowy and great egrets, and occasionally you'll spot animals like otter, mink and nutria.

Be sure to bring your binoculars and field guide to the low observation platforms at **North Pond dikes.** To catch glimpses of heron, egret and ibis rookeries, take the four-mile walk around North Pond. You'll notice sand fences built many years ago to stabilize the dunes, dikes around ponds and freshwater marshes planted for feed. Man has tampered with this land much to the delight of the winged ones which fly in without so much as an advance reservation and with no thought of check-out time.

The refuge was pretty well isolated until the bridge was built over Oregon Inlet in 1964. Then a few thousand visitors a year mushroomed to more than a million. They are rarely disappointed. From April to September diamond-back terrapins and yellow-bellied sliders sun themselves on pond banks, in September and October millions of Monarch butterflies pass through on their annual migration and November signals the peak of the swan migration.

Pea Island Refuge headquarters, approximately seven miles south of Oregon Inlet, provides visitors with folders on the birds most likely to be seen and gives current suggestions as to the best lookout spots. (8-4 daily. Refuge Manager, Pea Island, Box 606, Manteo, N.C. 27954. 919-987-2394.)

Chicamacomico Coast Guard Station, Hatteras (Photo by J. Foster Scott)

North Carolina

Hatteras Island

Chicamacomico Lifesaving Station, just south of Rodanthe on the ocean side, is where the National Park Service reenacts lifesaving operations with reproductions of equipment used at the turn of the century. (Thursday, 2 P.M. summer, free.) More than 600 ships have been wrecked in the past 400 years in the dangerous waters off Hatteras Island. No wonder the area is known as the graveyard of the Atlantic! In the 1870s the U.S. Lifesaving Service came to the rescue, building a chain of seven stations seven miles apart. Each station had horses, beach carts, boats and special lifesaving apparatus. When a ship was seen in distress the crew from the nearest station hauled the boat down the beach, launched into the turbulent surf and rowed to the foundering ship. Many people, otherwise doomed to a watery grave, were saved.

The U.S. Lifesaving Service played a vital role in Outer Banks history until 1915 when the Service merged with the Revenue Cutter Service to form the U.S. Coast Guard. Three modern Coast Guard stations now carry on the lifesaving tradition. The Chicamacomico Historical Association plans to restore this station since it was one of the first built on the Outer Banks. Eventually they hope to establish a national museum of the U.S. Lifesaving Service as a memorial to the heroic men who risked their lives for others. (Chicamacomico Historical Association, Inc., P.O. Box 140, Rodanthe, N.C. 27968.)

To reach **Cape Hatteras Lighthouse and Visitor Center** turn left from Highway N.C. 12 at Buxton. At 198 feet (208 if you measure from the base of the foundation to the peak of the roof), Hatteras

Cape Hatteras Lighthouse, Buxton (Photo by J. Foster Scott)

Sea Islands of the South

is the tallest lighthouse on the American coast. Its attractive spiraling black and white stripes make it one of the most photographed. It is certainly one of the most necessary! Hatteras earned its formidable reputation as the "Graveyard of the Atlantic" because the northern-flowing Gulf Stream comes perilously close to the Cape squeezing the southern current into a narrow passage around Diamond Shoals. These submerged fingers of sand are always shifting and passage through them is extremely tricky. Also, the Cape is the focal point for wild weather, "nor' easters" in winter, and tropical storms and hurricanes in late summer and early fall. Some ships which went down off this point of land disappeared without a trace, others left wrecks visible at low tide.

The original tower, built in 1803, had a small lamp fueled by sperm whale oil that didn't penetrate the darkness beyond the shoals. Storms sometimes shattered the windows putting the light out for days at a time. After a Fresnel lens was installed it gained a reputation as one of the most dependable lights on the coast. During the Civil War Confederates absconded with the lens in an attempt to confuse Union ships. When erosion nibbled at its base, the decision was made to erect a new tower 600 feet north of the original. The present brick structure was build in 1869-70 without the help of a modern pile driver. Two layers of six-by-twelve-inch yellow pine timbers placed crossways below the water table provided a floating foundation. Because the beams have always been submerged, they still are strong with no sign of rot or any other deterioration. The present light, installed in 1972, produces a beam of 800,000 candlepower visible for about 20 miles in clear weather. Why not climb the 275 steps and treat yourself to an extraordinary view?

The former keeper's residence (built in 1855) has been turned into a museum of the sea run by the National Park Service. Exhibits include lighthouse history, various rescue methods for saving shipwrecked souls and other aspects of man's relationship to the sea. Slide shows are regularly scheduled and a program of talks and activities offers something of interest to everyone. Choose from half-day hikes, fishing with a ranger, learning to surf or snorkle, sessions on Hatteras history or tales of dramatic shipwreck rescues. There's a nice selection of regional interest books, too. (Museum of the Sea, mid-June to Labor Day, 9-6, rest of year, 9-5. 919-995-5209.)

Near the Museum of the Sea Visitor Center is the **Buxton Woods Nature Trail** which is well worth your time. The trail, less than a mile, is in the only extensive forested area in the seashore. You'll come away with a much better understanding of the complex ecology of the Outer Banks environment, the interdependence of dunes, woods and marsh. You'll also understand why the Park Service makes such a point of asking visitors to use boardwalks to get

North Carolina

to the beach when you see the important role played by sea oats and other plants in stabilizing dunes.

Hatteras Village, the southernmost community on Hatteras Island, is the jumping off place for the free ferry to Ocracoke. This is a fishing village, pure and simple, said to be founded by shipwrecked sailors from Devon, England. The local accent is still flavored with Devon speech patterns.

Camping at the National Seashore is offered on a first-come, first-served basis. All campgrounds in the park have cold showers, drinking water, tables, grills and modern restrooms except Ocracoke which has pit toilets. There aren't any utility connections and remember that beach conditions require longer than normal tent stakes. (Division of Interpretation, Cape Hatteras National Seashore, Rte. 1, Box 675, Manteo, N.C. 27954. 919-473-2111.)

Hatteras Inlet Ferry To Ocracoke. Take a few crusts of bread to throw to the laughing gulls that hover behind the ferry to Ocracoke. The pleasant 40-minute trip runs fairly frequently in the summer time. Reservations aren't necessary, but you should be in the loading lane a half hour before departure if you're on a serious schedule. (Call 919-726-6446 or 726-6413 for information.) You may be interested to know the inlet you're crossing is relatively new. In 1719 shifting sands began to close the Hatteras Inlet and Ocracoke and Hatteras became one island. An 1846 hurricane reopened a new inlet making Hatteras eight miles shorter than in 1719.

Ocracoke Island

If you haven't seen Ocracoke, you haven't seen the Outer Banks! This 14-mile-long island (a half to two miles wide) still enjoys the relative isolation that comes from being reached only by water. This is, of course, part of its charm.

If you're approaching by ferry from the south rather than from Hatteras Island, remember that the summertime situation is crowded. Reservations are suggested to avoid long delays. They may be made within 30 days by phone or in person at the terminal of departure. Remember reservations are cancelled if your vehicle isn't in line a half hour before departure.

The name Ocracoke seems to inspire many a legend. The tale of Blackbeard raising a fist and yelling "Oh, Crow Cock!" to hasten the morning light seems a bit far-fetched. Early maps show a variety of Indian names for the island evolving from Woccocon to Occococ to Occocock which is a less romantic but more plausible explanation. Besides, the Indians were nicer.

The earliest European visitors were members of Sir Walter Raleigh's colony whose ship *Tiger* ran aground in Ocracoke Inlet in

Sea Islands of the South

1585. They landed briefly to make repairs. The original 17th century settlers were also of British descent. Some were shipwreck survivors, others came escaping the strict laws of the colony of Virginia. All developed the independent spirit that comes from being separated from the mainland and dealing with the capricious sea.

Their hazardous existence was made even more dangerous by the presence of pirates. In fact Ocracoke was headquarters for Edward Teach, better known as Blackbeard. The island inlet "Teach's Hole" was named for this unscrupulous character. Blackbeard ended up paying dearly for his barbarism. The infamous pirate was slain in a bloody battle in 1718 and his decapitated head mounted on the conquering ship's bowsprit, so people along the coast would know the reign of terror was over. Blackbeard's death marked the end of the Golden Age of Piracy in North Carolina.

The 1730 Village of Ocracoke was a tiny cluster of homes nestled in a protective hammock of trees on the sound side of the island. Soon Ocracoke Inlet became a major port of entry for ships coming to the colony. During the Revolution the treacherous waters of the Outer Banks worked in favor of the colonists by hindering British warships from adequately guarding the inlets. Desperately needed supplies for Washington's army at Valley Forge were successfully smuggled through the British blockade enabling the soldiers to survive their ordeal. But when shifting shoals began to block the inlet in the 19th century, Ocracoke's importance as a port declined drastically.

The islanders' isolation ended abruptly with the outbreak of World War II when a naval base was built on newly dredged Silver Lake. Five hundred men were stationed here. For the first time roads were paved and public phones installed. German submarines cruising near Diamond Shoals found merchant ships an easy target, and villagers watched in horror as American ships burned on the horizon. Coast Guardsmen patrolling on horseback found shipwreck debris and dead bodies on the island's beaches.

Four bodies that washed ashore were from the *H.M.S. Bedfordshire*, a British trawler patrolling the shipping lanes. Victims of a German torpedo, the officer and his three crew members were given a proper burial by villagers. During the 1976 Bicentennial, the State of North Carolina bought the grave site from private owners and presented the deed to England.

The 20th century arrived on this small island gradually. Private phones were installed and the state paved Highway 12 in 1957. Commercial fishing began to decline, as it has all along the Banks, but tourism grew. When Hatteras National Seashore was established in 1953, vacationing families joined the sports fishermen and hunters who had had the island to themselves for so many years.

North Carolina

The National Park Service preserved 5000 acres of the island including 16 miles of beach. It also took custody of the small herd of wild ponies which had roamed the island for as long as anyone could remember. Some think the ponies' ancestors came with the colonists or perhaps with early Spanish raiders. Others surmise they were survivors of shipwrecks.

What do you do on an unspoiled island? First, you take off your watch. Then listen to the wind. Watch the sandpipers play while you wait for the ebb tide to expose the old oyster beds. Sit on a Silver Lake dock until the sun sets. Fill your lungs with salt air. Sleep well.

When you're all loafed out, go talk to the famous Ocracoke ponies (if you can get that close). Or photograph the lighthouse. It's more than just a picturesque tower. The first light was built in 1798, the present one in 1823. During World War II a red lamp covered the beam in an attempt to confuse German submarines. Over the years it has served as a landmark for returning fishermen and a refuge during the violent storms. Since it was built on one of the highest spots in the village, the lighthouse shelters islanders when their homes are flooded. One senior citizen, whose parents were keepers of the light, remembers the night rising waters from a hurricane drove 75 people (and their pigs!) to their home. Her mother fed them all by putting lima beans on all four burners of the stove.

One good way to get acquainted with the village is to take the narrated historical tour on the **Ocracoke Trolley**. The one-hour open-air ride starts at Trolley Stop One (where you can also buy a hamburger and some yaupon tea) on Highway 12. You'll visit the

Ocracoke Harbor, Ocracoke, N.C. (Photo by J. Foster Scott)

Sea Islands of the South

graveyard where the British sailors rest and learn about 400 years of island history. (Wed. 10:30, 3, 7. Sunday, noon and 6, rest of week, 10:30, noon and 7 P.M. Trolley Stop One on Highway 12, 919-928-4041.)

The **National Park Service Visitor Center** is hiding out in one of the buildings used by the Navy during the war near the ferry slips for Cedar Island and Swan Quarter. Pick up information on the Discovery Adventures offered by the Park Service. They conduct day and night walking tours of the village, surfing lessons, pirate lectures and knot tying, to name just a few. (919-928-4531).

Ocracoke is a mecca for sportsfishermen. Surf casters reel in channel bass, puppy drum, blues and mackerel, flounder, grey and spotted trout, whiting and pompano. Boat fishing in the sound or the inlet for channel bass is best from April 15 to June 15 and Sept. 20 to Nov. 20. For cobia the best months are June and July, for blues and mackerel, July and August. One good thing about Ocracoke waters—something is always biting.

Guide service is available for fishing and hunting. For fishing call Captain Miller (919-928-5711), Captain Roberson (928-3361) or Captain Williams (928-3161). For hunting and fishing guide service get in touch with Captain Gaskill (928-3851) or Captain Isbrecht (928-5571). If you'd like to island hop to Beacon, Shell Castle and Pelican Island or cruise over to the ghost town on Portsmouth Island, call A.O. Burrus (928-3471).

CAPE LOOKOUT NATIONAL SEASHORE

Established in 1966, Cape Lookout National Seashore lies southwest of Ocracoke Inlet on three barrier islands—Portsmouth Island, Core Banks and Shackleford Banks. This narrow, 58-mile stretch of sand is being preserved in its natural state. Lovers of sea, sand and sun will enjoy day trips to this wild and lonely beach, as will campers used to primitive conditions.

Portsmouth Island

How to get there
Boat: Make arrangements on Ocracoke Island.
   Captain's names and numbers listed under Ocracoke Island.

Activites and accommodations
Swimming: Ocean (dangerous currents at times).
Fishing

North Carolina

Sightseeing: Ghost town of Portsmouth Village.
Camping: Primitive.
Accommodations: None unless you camp.

For more information
Cape Lookout National Seashore, P.O. Box 690, 415 Front St., Beaufort, N.C. 28516, 919-728-2121.

Portsmouth Village, once the largest community on the Outer Banks, is a ghost town. Settled around 1700 by pilots who guided ships into the inlet, this historic village is on the National Register of Historic Places. Ships arriving here from across the ocean were stopped by the shallow water between the island and the mainland, so they transferred their cargo into lighter shallow-draft boats for the last leg of the trip. To handle the loading, unloading and storage, men were hired and warehouses constructed. From 1735 on Portsmouth grew rapidly, peaking just before the Civil War with a population of 600.

The village was evacuated during the war except, the story goes, for one woman who was too fat to get through the doorway. (This is puzzling. Didn't anyone have an axe?) Those few who returned to their home town after the war reported her still there and quite satisfied with the way she had been treated by Northern soldiers.

The war was a terrible economic blow to the community but worse blows (literally) were to come. Storms filled in the channel and shipping activity, which had greatly decreased, ceased completely. After regular mail service ended, one of the last residents used to pole his skiff out to meet the mail boat which sailed between Cedar Island and Ocracoke.

Walk through the town and imagine a bustling port community. The citizens spent their leisure hours swimming, crabbing, clamming, and reading library books which arrived by boat. Croquet was a favorite pastime and for real excitement, there was an occasional square dance.

Visitors are warned to stay out of structures which may be unsafe or occupied by holders of National Park Service leases. You're welcome in the church. The beach is a mile walk from the village. Notice the red cedar, yaupon holly, wax myrtle and loblolly pine. Strong currents, rip tides and no lifeguard call for caution in the water.

Be sure to bring insect repellent, food, water and suntan lotion.

Sea Islands of the South

Core Banks and Shackleford Banks

How to get there
Ferry: Take U.S. 70, turn at State Road 1332 to Harkers Island ferry.
Private boat.

Activities and accommodations
Swimming
Fishing
Shelling
Camping: Primitive.
Accommodations: None.

For more information
Cape Lookout National Seashore, P.O. Box 690, 415 Front St., Beaufort, N.C. 28516, 919-728-2121.

Concessionaire ferry services are available from Harkers Island to the Cape Lookout Bight area. Since schedules are subject to change, ask the National Seashore agents for a current listing.

Most visitors to the Cape Lookout area are perfectly content to stroll the beach and gather shells or just lie back and watch the clouds. Those eager for more knowledge about the shore will enjoy a 15-minute orientation slide show given at **Hampton Mariners Museum** near the Beaufort waterfront. You might want to stop in at the museum or around the corner at the National Seashore headquarters for a complete list of offerings. Rangers at Cape Lookout give regularly scheduled talks on lighthouse history and conduct walks to introduce the island's plants, animals and shells. They even teach the fine art of fishing.

Most of the Core Banks is less than 2100 feet wide and just a few feet above high tide level. The most extensive forest, which consists of live oak, cedar, American holly, American hornbeam, Herculesclub and loblolly pine, is at Shackleford. But most of the shore is too low and narrow for stabilized vegetation. The seascape is mainly wide, bare beaches and low dunes covered with scattered grasses. Winds and waves rule, and sometimes the results are dramatic changes of land formation. You'll see weird "ghost forests" on the ocean side of some of the groves where sand and salt spray have killed the trees.

Bird watchers find the area rich in both variety and numbers. Large colonies of gulls and terns nest here and on several occasions an arctic bird, the razorbilled auk, has been sighted. Loggerhead

North Carolina

Cape Lookout Lighthouse

sea turtles lumber ashore to dig nests at the base of the dunes on summer nights and wild ponies, goats, cows and sheep roam Shackleford Banks.

Cape Lookout is labeled "Promontorium tremendum" on a 1590 map which means "Horrible headland." The treacherous shoals offshore spell danger, but the protected harbor is known as one of the safest on the entire coast. It was a refuge used by pirates, Spanish privateers and British warships in the Revolution. Union forces took the area over during the Civil War, and the harbor was a rendezvous used by the blockading fleet. Allied convoys during World War II sought shelter here, particularly after the antisubmarine net had been stretched across Beaufort Inlet.

Whaling began rather haphazardly in the 17th century with islanders depending on an occasional beached whale. They eventually mastered techniques for hunting the great mammals, but the industry died toward the end of the 19th century, and fishing took over as the economic mainstay. Diamond City on Shackleford Banks developed quite a reputation for the excellent salted mullet it exported. This small community of 500 residents was permanently evacuated when a hurricane flooded the town in 1899.

The **Cape Lookout Lighthouse** was completed in 1859 and 14 years later the distinctive black and white diamond pattern was painted on it to increase its effectiveness as a daymark. Although the lighthouse has survived hurricanes and war, it is now threatened by erosion.

Sea Islands of the South

The National Park Service has plans for additional visitor services which will include water, sanitation facilities, boat docks, camping areas and wayside exhibits. However, at the present time it's a good idea to bring everything you need with you.

Bogue Banks
Atlantic Beach and Emerald Isle

How to get there
Car: South of Morehead City, N.C., take U.S. 70 to the eastern end of the island, N.C. 58 and N.C. 24 to the western end.
Plane: Private. Beaufort/Morehead City airport.
Boat: Intracoastal Waterway.

Activities and accommodations
Golf
Tennis
Biking
Swimming: Ocean and pools.
Fishing
Boating
Children's activities: Programs conducted by Marine Resources Center.
Sightseeing: Fort Macon.
Camping
Accommodations: Rental cottages, motels, apartments, condominiums, hotels.

For more information
Carteret County Chamber of Commerce, P.O. Box 1198, Highway 70 West, Morehead City, N.C. 28557, 919-726-6831, Toll free in North Carolina 1-800-682-3934.

Bogue Banks, just south of Morehead City, North Carolina, is better known by the names of its communities, **Atlantic Beach** and **Emerald Isle**. Besides the lure of sun, beach and ocean, Bogue Banks offers Fort Macon State Park and a fine Marine Resources Center with an intriguing list of programs. Seafood restaurants, both in the fishing village of Salter Path on the island and in nearby Morehead City, serve delicious meals in a casual atmosphere.

Fort Macon, on the eastern tip of Bogue Banks, is an outstanding example of a 19th century military fortification. Early colonists in Beaufort were vulnerable to pirate attacks from the sea, but it wasn't until the Spanish attacked in 1747 that money was appropriated to build a number of forts. One, the victim of erosion,

North Carolina

has completely disappeared. Fort Macon was completed in 1826 and named for North Carolina's senator, Nathaniel Macon.

The state seized the fort from the federal government at the start of the Civil War but a year later Union troops captured it. Fort Macon was used as an important coaling station for Union ships and, after the war, as a federal prison. Although it was closed in 1877 when the United States adopted a policy of command of the seas rather than depending on coastal defense, it was garrisoned during the Spanish American War and World War II. In 1924 Congress deeded the fort to North Carolina which turned it into a state park.

If old forts don't appeal to you, think again. This is a particularly fine one with a small museum and an outstanding view from the parapets. Even if you don't tour the fort (which is free), you'll want to take advantage of the park recreation facilities — picnic tables, bathhouses and one of the finest beaches in the area. Local residents, who like to swim, surf, fish or picnic with an ocean view, congregate here.

Atlantic Beach, one of the state's oldest ocean resorts, has a well-developed amusement and commercial area. Sections of the beach have been set aside for surfing, and the protected waters of the

Fort Macon State Park

Sea Islands of the South

sound are ideal for waterskiing and boating. Fishing facilities include piers, two large headboats, charter boats and skiff rentals.

Salter Path, nine miles west of Atlantic Beach, is an old fishing village where people once lived an isolated existence entirely dependent on the sea and each other. Residents migrated here from ghost towns on Shackleford Banks or from other locations on the island which had been inundated by drifting dunes. Because they live on land which they never actually purchased, their homes must be handed down from generation to generation and occupied by the descendants forever or the family loses all claim to the property.

The **Theodore Roosevelt Natural Area,** five miles west of Atlantic Beach, is a favorite haunt of nature lovers and wildlife photographers. Its complicated ecosystem consists of salt marsh, inland freshwater slough and ancient dunes stabilized by dense maritime forests. The marsh is home to raccoons, fiddler crabs and birds — common egrets, green herons, clapper rails and osprey — and the sloughs are a haven for alligators. This is one of the few areas on Bogue Banks which has been saved from heavy development and its resulting erosion.

The North Carolina Marine Resources Center, located off Salter Path Road in the Roosevelt Natural Area, is a must. Attractively designed displays focus on the marine world, the fishing industry and the ever-changing geography of a barrier island. The touch tank, with marine life accessible for handling, intrigues children as do the turtles and fish in the aquaria. Wide-ranging offerings include films *(Dangerous Animals of the Sea),* demonstrations (Japanese art of fish printing), talks ("Why Won't Barrier Islands Stand Still?") and field trips to the beach and marsh. (Open Mon.-Fri., 9-5, Sat. 10-4 and Sun. 1-5 year 'round. 919-726-0121.)

Emerald Isle which makes up the west half of Bogue Banks is a beach community where cottage owners and renters look to the sea for its multiple pleasures. The view from the bridge between Emerald Isle and Cape Carteret with myriad marshy inlets dotting the intracoastal waterway is nothing short of splendid.

Bear Island
Hammocks Beach State Park

How to get there
Car: Five miles southeast of Swansboro via N.C. 24 to the free passenger ferry.
Boat: Private boats may be tied up at the ferry dock on the island, or they may be beached along the shore.

North Carolina

Activities and accommodations
Swimming: Ocean.
Fishing
Camping: Primitive.
Accommodations: None.

For more information
Hammocks Beach, Rte. 2, Box 136-B, Swansboro, N.C. 28584.

Bear Island, accessible only by water, is five miles southeast of Swansboro, North Carolina, between Beaufort and Wilmington. This pristine bit of wilderness has a beautiful wide sandy beach and spectacular dunes, some as high as 60 feet. If you have lost faith in the world as a beautiful place, you'll regain it here. Bear Island should have been mentioned in that old song, "The Best Things in Life are Free," because there's no charge at all, even for the boat ride. This is definitely one of the best bargains on the coast!

The small 892-acre island with three miles of splendid white beach is crowd-free. The park, which includes the entire island, has picnic tables, a bathhouse with restrooms, showers, refreshment stand, and lifeguard a half-mile walk from the ferry landing. Visitors to the island swim, surf, look for shells, hike, fish (channel and surf) or simply soak up the sun.

Two free passenger ferries operate between the mainland and Hammocks Beach. They run daily from Memorial Day to Labor Day making one round trip each hour. The first trip leaves the mainland at 9:30 A.M.; the last ferry departs the island at 6:00 P.M. In May and September the ferry operates only on weekends. The two-and-a-half-mile trip, which meanders through the marshy islands dotting the Inland Waterway, takes approximately 25 minutes.

You'll have to share the island with the birds — laughing gulls, terns and oyster catchers — who think they own the place. Loggerhead sea turtles emerge on summer nights to make their nests above the the high water mark, and ghost crabs scuttle sideways across the sunbaked sand.

In 1713, according to the Colonial Records of North Carolina, Bear Island was acquired by Tobias Knight who was said to have pirate connections. Rumors of buried treasure are still around! The colonial assembly built a fort here shortly after a Spanish attack, but until the early 1900s Bear Island was used only sporadically by whalers, fishermen and a nudist colony.

In 1914 Dr. William Sharpe, a pioneer in brain surgery, joined some colleagues on a duck hunting expedition to North Carolina. He ended up in a small boat guided by John Hurst, an expert outdoorsman. The friendship that developed from that first meeting was to dramatically affect the island's future.

Sea Islands of the South

Dr. Sharpe asked John Hurst to find him land in the area, "a retreat which would be beautiful, isolated, and have an abundance of fish and game." After a three-year search, Hurst came up with the Hammocks, a large section of the mainland peninsula which included the small island. William Sharpe had been reared in the slums of Pittsburgh, Chicago and South Philadelphia. No wonder he fell in love with this area which he described as a place "where every day is Sunday, where the climate is wonderful the year round."

In his autobiography, *Brain Surgeon,* Sharpe wrote of his 38-year friendship with John Hurst: "I realize I have gained from him more off-the-beaten track, common-sense, useful bits of knowledge than from any other man, not only because he is a master at farming, stockraising and outdoor life, but because he is a philosopher. To him, the good luck to have been born in itself compensates for the many tragedies of life."

Dr. Sharpe stirred quite a local controversy when he hired John to manage his land. He received an unsigned letter objecting to the black manager, stating that a white should have the job and threatening damage to the property. Sharpe silenced the threats by inserting a notice in the local paper offering a $5000 reward for information leading to the arrest and conviction of anyone injuring the personnel or causing damage to the island.

Dr. Sharpe planned to will the island to Mr. and Mrs. Hurst, but they persuaded him to give it to the black teachers of the state instead. The island was used as a meeting place and recreation center for the teachers, and black children constructed a 4-H camp and Future Farmers of America camp. After the schools were integrated, the black teachers no longer existed as a separate group and financial problems developed. In 1961 the teachers presented Bear Island to the state.

It's hard to believe this exquisite jewel is no longer the sole province of the very wealthy or of a special group but a resource for everyone.

Topsail Island

How to get there
Car: Five miles east of U.S. 17. Access is by drawbridge on N.C. 50 at Surf City or by span bridge on N.C. 210 near the north end of the island.

Activities and accommodations
Golf
Tennis
Biking

North Carolina

Swimming: Pools and ocean.
Fishing: Head boats, charter boats, surf and pier fishing.
Boating
Camping
Accommodations: Rental cottages, hotels, motels, apartments.

For more information
The Greater Topsail Area Chamber of Commerce, P.O. Box 486, Surf City, N.C. 28445.

Topsail Island, with 26 miles of surf-washed beach, is located between Wilmington and Swansboro, North Carolina. Bounded by the New River Inlet to the north, Topsail Inlet to the south, backed by the Intracoastal Waterway and facing the Atlantic Ocean, the island is a refuge for water lovers of all ages. Swimming, boating, diving, fishing and beachcombing lure some while others find their relaxation contemplating the ocean's many moods.

Topsail is primarily a family cottage affair with spots of commercial development, mainly in the form of small shopping centers, in Surf City and the town of Topsail Beach. Accommodations range from weathered cottages to spiffy motels.

Sea Islands of the South

The island provided perfect cover for pirates in the early 1700s who hid their ships in the channels behind the dunes. When a merchant ship appeared on the horizon, the pirates would take off after it, loot the vessel and kill the crew. Captains of merchant ships were understandably nervous passing the island and would post watches for the tops of sails showing above the dunes. The island has been known as Topsail ever since.

Salt was made on these shores until the Civil War when Union troops went to a great deal of trouble to destroy the salt works. Until World War II access was by boat only. In 1940 the Government developed a part of the island in conjunction with Camp Davis on the mainland. The peaceful beach became a bustling community with the addition of a pontoon bridge, roads, fresh water, power lines and the construction of a number of buildings. All this could hardly compare with the activity which was to come later when the government chose the area for its missile launching project. The concrete towers and launching pads were later abandoned when the project moved to Cape Canaveral.

The island was released to land owners in 1949 and, in a short time, was converted to the ocean resort which year after year draws vacationers in quest of sun, sand and surf.

Figure Eight Island

How to get there
Eight miles north of Wilmington (just north of Wrightsville Beach), two miles east of Highway 17, across private security-guarded drawbridge.

Activities and accommodations
Tennis
Biking
Swimming
Fishing
Boating and sailing
Sightseeing: Historic Wilmington.
Accommodations: Private residential island limited to homeowners and their guests.

For more information
Figure Eight Island, 15 Bridge Road, Wilmington, N.C. 28401, 919-686-0631.

Figure Eight is a strictly private island of green highlands and rolling dunes plumed with breeze-tossed sea oats. Its five miles of beach and wide stretches of salt marsh are separated from the

North Carolina

mainland by the Intracoastal Waterway. The blue Atlantic serves up great doses of seaside serenity to the fortunate residents of this secluded community.

The primary theme of Figure Eight's development has been preservation of the island's natural beauty. More than 600 lots have been situated so that every home is either on the sound or ocean waterfront. Approximately 150 residences, each of an original design and approved by a review board, have been carefully integrated into the fragile beach environment. The yacht club, currently the only non-residential building on Figure Eight, recently won an award for architectural excellence, as have many of the island homes.

According to legend, the island was named by pilots who felt they were sailing in a figure 8 as they navigated the old winding inland waterway channel. Originally owned by James Moore, Figure Eight passed through a succession of private owners until it was purchased in the mid-Fifties by two brothers who initiated its development. Today's affluent homeowners share this sunny stretch of sand with a host of wild inhabitants whose claim to the island is aboriginal.

Wrightsville Beach

How to get there
Nine miles east of Wilmington, N.C. via Rtes. 74 or 76.

Activities and accommodations
Tennis
Golf: Nearby.
Biking
Swimming
Fishing
Boating
Waterskiing
Sightseeing: Historic Wilmington.
Accommodations: Hotel, motels, condominiums, rental cottages, apartments.

For more information
Greater Wilmington Chamber of Commerce, P.O. Box 330, Wilmington, N.C. 28401, 919-762-2611.

Looking for a well-developed resort area where you can swim, waterski, fish or sail near city conveniences? Consider Wrightsville Beach just east of historic Wilmington, North Carolina. Here you can suit your budget and lifestyle with a variety of accommodations

Sea Islands of the South

ranging from modest apartments to luxurious condominiums or high rise hotel rooms. You may fish in the surf, from a pier, or book passage on a charter boat. After you've landed the big ones, reward yourself with a night on the town. A melange of superb restaurants, both in Wrightsville Beach and nearby Wilmington, cater to vacationers with discriminating palates. Sheltered soundside waters provide ideal conditions for sailboat races and waterskiing, and the rhythm of the surf guarantees a good night's sleep.

Pleasure Island
Carolina Beach and Kure Beach

How to get there
South of Wilmington on Rte. 421.

Activities and accommodations
Tennis
Biking
Swimming
Fishing
Boating
Sightseeing: Fort Fisher, Blockade Runner Museum and nearby historic Wilmington.
Camping
Accommodations: Motels, hotels, condominiums, rental cottages.

For more information
Pleasure Island Tourist Bureau, P.O. Drawer A, Carolina Beach, N.C. 28428, 919-458-8434.

Sun, sand and surf. Boardwalk amusements, rides, games, movies and shops. An historic fort, museums and aquaria. Memorable meals in seafood restaurants. Shady state park campgrounds and condominiums. Who says you can't have it all?

Pleasure Island, better known as Carolina and Kure Beaches, is just south of historic Wilmington, North Carolina. If, like most island lovers, your interest centers on the sea, you'll enjoy browsing among the attractions here.

Start with the **Blockade Runner Museum** near Carolina Beach. You'll learn all about the exploits of the brave men who dared slip their sleek vessels past Union ships to bring precious supplies to Wilmington. With running lights doused and fearful of going aground on the shoals, they lived in constant danger of being caught and killed. A dramatic audiovisual presentation, scale-model waterfront scene and artifacts retrieved from the ocean floor

North Carolina

tell the story. (Charge. Tues.-Sat. 9-5, Sunday 1-5, 919-458-5746.)

Nearby **Carolina Beach State Park** with its new marina is a favorite with boaters as well as tent and trailer campers. The pretty setting overlooks the Cape Fear River where the fishing is action-packed. Of the several unusual varieties of coastal flora found here, the unique and rare Venus flytrap gets the most attention. Did you know this insect-eating plant, which Darwin called the "most wonderful in the world," grows nowhere else but in the Carolinas?

If you enjoyed the Blockade Runner Museum, you'll want to know the rest of the story. **Fort Fisher,** south of Carolina Beach, is the bastion which kept the Cape Fear River open to blockade runners during the most of the Civil War. Until January 1865 Confederate armies' supplies came through Wilmington. Fisher was the Civil War's largest earthworks fort and it took the most extensive land-sea battle prior to World War I to bring it to its knees. The Cape Fear region had been a keystone of the Confederate war effort and when the fort fell, the last significant source of overseas supply was blocked. The South lasted only 90 days after Fort Fisher's surrender.

Visitors are welcome to walk the well-kept (and sometimes mosquitoey) grounds, see the slide show that tells the fort's story and inspect artifacts recovered from the blockade running era. (Free. Tuesday-Saturday, 9-5, Sunday, 1-5.)

Don't miss the **North Carolina Marine Resources Center** at Kure

N.C. Marine Resources Center

Sea Islands of the South

Beach where changing exhibits probe the mysteries of the seas. Here you'll learn about the formation of islands, the devastating force of a hurricane and the best way to combat erosion. You may pick up live starfish and sand dollars in the touch tank or just observe native fish and turtles in the aquaria. The center is roomy and well laid out, a real find for vacationing families. (Free. Mon.-Sat. 9-4:30, Sun. 1-4:30.)

Want to continue southward? Line your car up for the Southport-Fort Fisher ferry. (Charge) You'll pass ocean freighters and shrimp boats on the one-hour trip. Don't forget the bread; those gulls haven't had anything to eat since the last ferry, or so they would have you believe. (Ferry Manager's Office, North Carolina Department of Transportation, Morehead City, N.C. 28557. 919-726-6446.)

Bald Head Island

How to get there
Boat: At the mouth of the Cape Fear River three and a half miles east of Southport, N.C.
Light plane: Airstrip.

Activities and accommodations
Golf
Tennis
Swimming: Ocean.
Fishing
Accommodations: Eight room inn.

For more information
Bald Head Island Corp., P.O. Box 1058, South port, N.C. 28461, 919-457-6763.

Bald Head Island, just off the southernmost tip of North Carolina's coast in the mouth of the Cape Fear River, is accessible only by water or air. This unique semitropical island is in the initial stages of development.

Bald Head is one of the most beautiful islands on the coast. Fourteen miles of sandy white shoreline encircle dunes, marshes full of birds and a dense maritime forest of palms, live oaks, cedar and dogwood. This is the northernmost island where palmettos thrive naturally, the only place in North Carolina. Fox, raccoon, opossum, squirrel and alligator have the run of Bald Head which serves as a wintering ground for migrating waterfowl. Loggerhead

North Carolina

sea turtles drag their heavy bodies ashore by the light of the summer moon to dig nests and lay their eggs.

To these riches developers have added an 18-hole championship golf course, tennis courts, an eight-room inn and restaurant. Plans call for an exclusive, environmentally oriented residential resort community. The emphasis is on seclusion, the tone is low key, the commitment is toward low density development. The majority of the island is to be left in its natural state.

Unless you own property or are interested in buying some, you probably won't get to see Bald Head unless you are a golfer. A special golf package which includes boat transportation to and from the island and a cart and greens fee is available from Wednesday through Sunday. The course has received rave reviews in golfing magazines and the island is a gem. This is the best reason we've heard of to take up golf!

Bald Head has had a colorful past. Several Indian sites have been discovered, French and Spanish explorers landed here in the 16th century, and the island was base of operations for a band of pirates in 1718. These buccaneers set decoy lights along the shore which lured merchant ships onto Frying Pan Shoals. When the ships ran aground, the pirates looted them and killed those aboard. In two months they captured 13 ships, but their reign of terror ended abruptly when they were caught and hung.

Three different lighthouses have helped guide ships through these dangerous waters, and members of the Lifesaving Service were once stationed here. Their weatherbeaten buildings are presently undergoing restoration. **Old Baldy**, the second lighthouse, is no longer used for navigation but still proudly oversees the island. Built in 1817, this octagonal structure is the oldest building on Bald Head and the oldest standing lighthouse in the state.

For a time the island was farmed, but wars disturbed its tranquil isolation. During the Revolution more than 5000 British soldiers were stationed here and the Confederacy built Fort Holmes on the southwestern side during the Civil War. Between the fort and the treacherous shoals, North Carolina was able to keep the Cape Fear from being totally blockaded.

Smith Island, as Bald Head was known and is still shown on some maps, was named for its first owner, Thomas Smith. His grandson, a North Carolina governor in the early 1800s, donated ten acres to the government for a lighthouse. Thomas Boyd of Wilmington had great plans when he bought the island for $4500 in 1914. He built a hotel and began an extensive resort development, but he was caught by the depression and ended up offering it to the state in 1933. The state never appropriated the funds, so it reverted to Brunswick County for taxes. From 1938 to 1963, Mr. and Mrs. Frank Sherrill purchased the island section by section from the county. They even bought the lighthouse site from the federal government. In 1970 the Carolina Cape Fear Corporation paid five and a half million dollars for Bald Head and began development, but a great controversy ensued. The issue of a permit to build a marina lined the conservationists up against the developers and for eight years the case was in and out of courts. Financial troubles again plagued the developers and the Bald Head Island Corporation took over in 1976.

The upshot of all the controversy was finally settled in court when the developers were granted permission to go ahead with marina construction. To allay the state's fears that the precious marshlands and other natural treasures of the island would be destroyed, the developers deeded 10,000 acres of waterways, marshes and uplands to the North Carolina Nature Conservancy. That left Bald Head Corporation with about 3000 acres, 1800 of which they proposed for development.

At the present time, although several hundred owners have lots, less than 25 have built homes on them. All proposed building plans have to be passed by an architectural review board. Houses must blend with the landscape and should not tower above the tree line. The developer has pledged to preserve as many of the island's natural features as possible.

North Carolina

Oak Island
Yaupon Beach, Long Beach and Caswell Beach

How to get there
Car: Exit off U.S. 17 onto N.C. Rtes. 133, 87 or 211.
Plane: Brunswick County Airport is located between Oak Island and Southport. Jetport at Wilmington, N.C., is a 30-minute drive.
Boat: Marine facilities on Intracoastal Waterway which separates Oak Island from the mainland.

Activities and accommodations
Golf: Oak Island Golf Club in Yaupon Beach has an 18-hole championship course. You may make arrangements at this club to play on the renowned course on Bald Head Island. The greens fee includes a round-trip ticket by boat from Southport.
Tennis
Biking
Swimming: Pools and ocean.
Fishing: Outstanding facilities at Southport.
Boating: Intracoastal Waterway.
Sightseeing: Historic attractions nearby.
Camping: Two family campgrounds near oceanfront in Long Beach.
Accommodations: Motels, rental cottages, apartments.

For more information
Southport-Oak Island Chamber of Commerce, Box 52, Southport, N.C. 28461, 919-457-6964.

Oak Island, about six miles from Southport, North Carolina, is better known by the names of its three separate beach communities— Yaupon, Long and Caswell. Bounded by the Atlantic Ocean and the Intracoastal Waterway, the island is reached by a new high-level bridge. Its 14 miles of sandy beach draw sun worshippers who find plenty to do without the heavy concentration of commercialism of some mainland beach communities.

Swimming and beachcombing are perennial favorites, but golf, tennis, waterskiing, biking and fishing are also popular. The Southport-Oak Island area has long been a year-round mecca for fishermen. Here you have a number of choices: pier fishing, surf casting or deep sea fishing in a charter boat. You can fish the Atlantic, the Intracoastal Waterway, the Cape Fear River and any one of several inlets, canals and estuaries. At Southport you'll see

Sea Islands of the South

boats unload catches of flounder, mackerel, mullet, sea bass, spot and whiting.

The island is primarily a family resort with emphasis on casual living. Rental cottages run the gamut from elegant to run down; something is always available. The chamber of commerce will send Southport's weekly newspaper on request free of charge, which gives information on properties available for rent.

Oak island comes by its name honestly; much of it is heavily wooded. However, some of the beach development has grown like topsy. You'll see erosion lapping at the edges of, or actually undercutting, some houses which were not planned with the shifting sands of a barrier island in mind.

South Brunswick Islands
Holden Beach, Sunset Beach, Ocean Isle Beach and Bird Island

How to get there
Car: These beaches are just off the coast at the southeastern tip of North Carolina approximately 30 miles south of Wilmington, N.C. and 30 miles north of Myrtle Beach, S.C. Holden, Sunset and Ocean Isle are east of Rte. 17 and reached by drawbridges across the Intracoastal Waterway. Bird Island is accessible only by boat or by wading across from Sunset Beach at low tide. (Watch this tide carefully and plan your return for daylight hours.)

Plane: By private plane: Ocean Isle Beach has a paved, lighted airstrip.

Activities and accommodations
Golf
Tennis
Biking
Swimming: Motel pools and ocean.
Fishing
Horseback riding: Nearby.
Boating
Sightseeing: Gardens and plantations a short drive away.
Camping
Accommodations: Motels, vacation apartments, cottages.

For more information
South Brunswick Islands Chamber of Commerce, P.O. Box 784, Shallotte, N.C. 28459, 919-754-6644.

North Carolina

These lovely beaches, half way between Myrtle Beach, South Carolina, and Wilmington, North Carolina, are tranquil havens for those who love the ocean. Holden, Sunset and Ocean Isle are family-style islands, served up without the fanfare of crowds and amusements. The beaches are lined with cottages for people who don't need to be entertained, whose pleasures come from sunrise and sunset, fresh seafood, clean salt air and a dip in the surf.

Bird Island is undeveloped and reached only by boat or by wading across from Sunset Beach at low tide. It's a fine place to look for shells or have a picnic, but don't say you weren't warned. Those who go without a tide table and a watch, or just some good old-fashioned common sense, have been known to spend a long, lonely night there waiting for the tide to go out.

Outdoor lovers appreciate the island's long freeze-free season which averages eight months. Thanks to the Gulf Stream, winter (December through February) is warmer here than in Phoenix, Arizona or Mexico City. In April the weather is comparable to Bermuda and during summer the sea breezes provide natural cooling.

Recreation centers around the ocean—swimming, sailing, boating, fishing and beach combing. The atmosphere is tranquil, the beaches wide and sloping, the sand white and powdery. If you want to join the herds of people at Myrtle Beach or explore historic Wilmington, you'll find those cities within 30 miles.

Fishing opportunities abound. Choose from fresh and brackish streams, saltwater rivers and creeks, and the Atlantic Ocean. If you picked the ocean, you have to decide whether to surf fish, charter a boat to the Gulf Stream or pier fish. There's a modest fee for pier fishing, but walking out on the pier to see what's biting is a time-honored custom that's free.

A buoy visible from Holden Beach marks one of three artificial fishing reefs in the county. These are made from old tires loaded with concrete, cabled together and dropped over the reef. Divers have substantiated that an artificial reef, whether it began with a sunken ship or a load of tires, attracts fish immediately. Barnacles, other crustaceans and moss attach to the reef material quickly. This provides food for the small fish and the small fish provide food for the big ones. The object, to bring good fishing in close to shore, is realized.

You're not interested in fishing but love seafood? You're in the right place. The emphasis here is on fresh rather than fancy. You may buy fresh shrimp, oysters and fish where the fishing boats tie up at Seaside Landing or try the island restaurants.

You may want to sample the fare in **Calabash**, the self-proclaimed "Seafood Capital of the World." Calabash is just a few minutes drive from the islands. It's less than a mile between city limit signs but within that space are more than 20 seafood restaurants. These independent establishments, with nary a golden

Sea Islands of the South

arch in the lot, have a far-flung reputation as a look at the license plates in their parking lots testifies. The town's seafood legacy goes back to the 1920s when shrimpers and other fishermen sold their catches and to the depression years when river folk traded clams and oysters for vegetables here. The restaurant business started when popular outdoor oyster roasts for tourists had to be moved inside because of health regulations.

If statistics impress you, think of this: The town, with a population of about 150, serves about one and a quarter million hungry people a year. We assume they were hungry because they consumed 378,000 pounds of shrimp and 668,000 pounds of flounder, not to mention the scallops and oysters. You won't need statistics to determine the popularity of the restaurants. All you have to do is join the line of cars that creeps through town around dinner time in the summer. If you're allergic to lines, try eating early or late. Don't dress up; the word here is casual.

With golf, tennis, ocean-related fun and all that good eating, you need only decide which island to visit. Holden Beach, at 11 miles, is the largest, Sunset the smallest and Ocean Isle in between the two is, appropriately, middle-sized. Holden is nearest to Wilmington and Sunset closest to Calabash and Myrtle Beach, but Ocean Isle people insist they wouldn't be anywhere else.

Shrimping Fleet

Islands of South Carolina

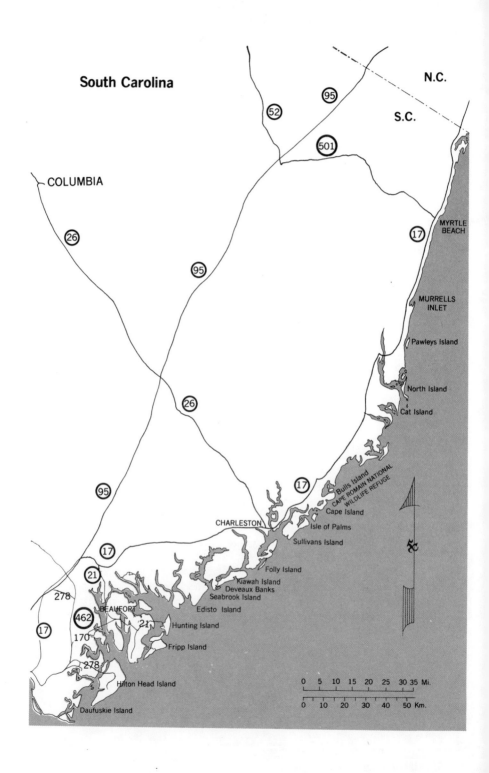

South Carolina

Pawleys Island

How to get there:
A 15-minute car ride north of Georgetown, or 25 miles south of Myrtle Beach, just east of Rte. 17.

Activities and accommodations
Biking
Swimming
Fishing
Crabbing
Loafing
Accommodations: Motels and rental cottages.

For more information:
Greater Myrtle Beach Chamber of Commerce, P.O. Box 2115, Myrtle Beach, S.C. 29577, 803-448-5135.

Pawleys Island, four miles long by a very slim fourth of a mile (at its widest point!) is a 15-minute car ride north of Georgetown, South Carolina. Twenty-five miles south of Myrtle Beach, the island is a world away from the clamor, lights and crowds of that popular resort. Devotees who know its personality and charm as so distinctive could never consider Pawleys just another island.

Besides picturesque old homes with gabled roofs and around-the-house porches, Pawleys is renowned for outstanding seafood, handwoven rope hammocks and its own resident ghost. Leave your pretensions on the mainland. When you cross the causeway to the island, you're just another sunset gazer whose major mission is to attune yourself to the rhythm of the tides.

Pawleys has been a well-loved vacation retreat since the late 1700s, making it one of the oldest beach resorts on the country's eastern coast. Wealthy Carolina planters and their physicians were convinced the raging fevers that claimed so many lives each summer were caused by the decay of vegetation. George Pawley II's descendants and approximately 100 other plantation families were the first to seek the island's salt air in order to escape from the deadly "country fever," as malaria was then known. They left their plantations, literally fleeing for their lives, in early May and returned only after the first hard frost, usually in November.

Indians are said to have started salt making on the island. The low country tribes—Seewee, Sampit, Winyaw, Peedee and Wac-

Sea Islands of the South

camaw—were the first to work out a way of evaporating sea water to get salt. In 1782 Percival Pawley was pumping sea water into shallow ten-foot-square vats and letting it evaporate in the sun. Salt produced at Pawleys was used by troops in the Revolutionary War, and records show that by 1862 more sophisticatd equipment was capable of producing between 30 and 40 bushels of salt a day. Production came to an abrupt halt in 1863 when Union troops completely demolished the salt works.

The Civil War was devastating to the fortunes of the plantation owners. Some tried to make a comeback after the war, but several tropical storms wrecked their crops, and they faced increasing competition from rice growers in Texas, Louisiana and Arkansas. Although the luxurious life of the plantation era was over, many of the planters managed to keep the island homes which meant so much to them.

Violent storms play a prominent role in all island histories but particularly in one a scant quarter mile wide. Fortunately, many islanders in the past have heeded the warning of the "Gray Man," the ghost sometimes seen when Pawleys was in danger of being hit by the full fury of a hurricane. Those who had not seen the ghost or understood the significance of his presence had some terrible tales to tell. Destruction was worst when the full force of the storm coincided with high tide. When that happened, houses were carried out to sea and whole families were torn, one by one, from tree limbs and were swallowed by the raging ocean.

Two never-to-be-forgotten storms that hit Pawleys were one in 1893 and Hurricane Hazel on October 15, 1954. Hazel destroyed

Beach fun (Photo courtesy of Kiawah Island)

South Carolina

practically every new house which had been built on the island's south end since World War II. Many of the old houses which had been constructed of heavy handhewn timber on solid foundations survived, but the high dunes which had protected them were leveled. Beach residents, having received a three-hour warning from the weather bureau, evacuated the island without a single fatality. Today the old houses are proud reminders of an era of fine craftsmanship when Pawleys was home for half the year, not a few fleeting summer weeks.

Inns and small motels cater to devoted Pawleys Island lovers who return year after year. The first thing vacationers do is to take off their shoes. The next order of business is to let the day unfold of its own accord. They swim, sun, fish, crab or visit the neighbors. Nightlife consists of steaming their own catch of crab or staking out a table at one of the nearby seafood restaurants where the shrimp is barely out of the net before it's cooked.

A favorite stop is the Hammock Shop on Rte. 17. There children can try out the joggling board before buying a treat from the summer house ice cream parlor. Native craftsmen carry on the tradition of making rope hammocks and a tinsmith turns out old-time lanterns while other artisans demonstrate their skills and answer questions. The shop, which calls itself the Home of the Original Pawleys Island Rope Hammock, features a staggering selection of international gourmet foods and an impressive array of needlework materials as well as books, prints, gifts and souvenirs.

Bulls Island
Cape Romain National Wildlife Refuge
Including Cape Island, Lighthouse Island, Raccoon Keys and Other Islands

How to get there
Car: Turn off Highway 17 about 20 miles northeast of Charleston onto SeeWee Road for 5 miles to Moore's Landing. Boat access to Bulls Island is generally available from this point.

Activities and accommodations
Fishing: Excellent!
Visitation: National Wildlife Refuge, daylight hours only.
Camping: None.

Sea Islands of the South

For more information
Refuge Manager, Rte. 1, Box 191, Awendaw, S.C. (803-928-3368).

Bulls Island in Cape Romain National Wildlife Refuge is about 20 mile northeast of Charleston, South Carolina. Boat transportation must be privately arranged as there are no regularly scheduled trips. A five-mile boat ride through tidal marshes and winding creeks takes you to Bulls Island, the showcase of this 20-mile long, 60,000-acre refuge.

Cape Romain is considered one of the premier wildlife sanctuaries in the East. Because it is fairly accessible and has more than 150 species of birds, Bulls Island is a favorite of bird lovers. The great flocks which winter here include gulls, cormorants, horned grebes, brown pelicans, black skimmers, peregrine falcons, black ducks, pintails, mallards, gadwalls, teals, Canada geese, ring-necked ducks, scaupa, buffleheads, shovelers, coots and common gallinules. Most of the Atlantic Coast population of American oyster catchers congregate in the refuge during the colder months, and the Cape is one of the last undisturbed nesting areas for the endangered eastern brown pelican. After the migrating birds have gone north, flocks of wood ibises, snowy egrets and herons arrive for the summer months.

While some of the islands in the refuge are so low they disappear during high tide, Bulls Island is elevated and covered with dense semitropical growth. Besides stands of bamboo, live oaks, magnolias, pines, laurel and palmettos, this six-by-two-mile island has over 20 miles of sandy trails, long shell-strewn beaches, huge dunes, water lily covered ponds and the ruins of an old fort. Bulls is an important nesting area for the endangered loggerhead sea turtle and boasts a thriving population of deer, wild turkey, raccoon, black fox squirrel and alligators.

This ancient barrier reef, often the first land sighted by new settlers coming to the South Carolina coast, was named for Stephen Bull, one of the leaders on the ship *Carolina*. All that remains of the fort where colonists and Indians watched for pirate ships in Bulls Bay are the ruins of an old tabby wall, made from a mixture of oyster shells, lime and sand.

Each season is special. The best time to see a great number and variety of birds is during the fall, in spring when the wood ducks nest and in winter when the ponds are full of thousands of waterfowl. The pelicans return in February, in April loggerhead sea turtles mate in the bays and tidal creeks, the alligators begin to nest in May, the turtles lumber ashore to lay their eggs by the light of the summer moon, in November the channel bass run is at its peak and whistling swans may be seen on Upper Summerhouse Pond in December.

South Carolina

Since only daylight use of the island is permitted, camping is not allowed. Two nearby campgrounds are Buck Hall in **Francis Marion National Forest** (ten miles north of the refuge) and privately-owned **Camp See Wee** (five miles north). Both are accessible from Rte. 17.

Take everything you need as only restrooms, drinking water and picnic tables are available. Repellent is strongly recommended as insects can be a problem, especially in the summer when the breeze dies. There is a blind for photography. Weapons and pets are not permitted. If you have questions, stop in the Refuge Headquarters at Moore's Landing.

Isle of Palms

How to get there
Car: From Charleston, take U.S. 17 N (business) to Mt. Pleasant, turn right on S.C. 703.

Activities and accommodations
Tennis
Biking: Rentals available.
Swimming: Pool and ocean.
Fishing: Charter boats nearby.
Boating: Rental sailboats available.
Children's activities: Well-supervised, fun program.
Accommodations: Rental cottages and villas.

For more information
Isle of Palms Beach and Racquet Club, P.O. Box Y, Charleston, S.C. 29402.

Isle of Palms, 15 miles from Charleston and eight miles from the village of Mt. Pleasant, is literally an Isle of Palms. The island's dense subtropical jungle is just one of its many natural gifts. Others include an uncrowded beach overflown by crowds of pelicans, a gentle surf and white dunes plumed with graceful sea oats. When you add this beauty to the proximity of Charleston with its historic charms and fine restaurants, you come up with a winning combination.

The Isle of Palms Beach and Racquet Club offers all the standard recreational facilities (with the exception of a golf course which is planned), a well-supervised kids' program and 24-hour security. Those who can't relax and enjoy themselves here ought to do some soul searching. If you need crowds a la Waikiki or dawn-to-dusk action beyond a fast-paced tennis game, you're in the wrong environment.

Sea Islands of the South

Photo courtesy of Charleston Chamber of Commerce

Long Island, a former name of the Isle of Palms, was the home of the Seewee Indians. Their simple life of hunting and growing vegetables was greatly complicated by the arrival of the Europeans. Soon the Indians were supplying both warriors and provisions to colonists under seige by the French and Spanish. Later, they traded skins and furs with the English, but thought they were being cheated. In order to cut out the middle men, they decided to go directly to the King of England with their goods.

They based their expedition on the conclusion that the King couldn't be far away since so many English ships came into the Charleston harbor. After a year making dugout canoes and accumulating furs to trade and provisions for the trip, they headed for England. At first the winds were brisk and the sails made from mats worked well. Unfortunately, a tropical hurricane finished off the adventure, and the few who were saved by passing ships were sold as slaves in the West Indies. Those left behind on the island, the young and the infirm, were wiped out by smallpox or absorbed into other tribes in the Catawba Nation. The Seewees were the first Carolina tribe to disappear completely.

After the American revolution, Charlestonians began to go to nearby Sullivans Island during the hot summer months, but since there was no access to the Isle of Palms, it remained a wilderness. The owner, after deciding to develop the island, became instrumental in bringing the Charleston and Seashore Railroad to it. This train carried its first passengers during the summer of 1898 and was, at first, wonderfully successful. Bathhouses, a restaurant, dancing pavilion and a grand hotel followed in short order. Charlestonians would take a trolley to the wharf, ride the ferry to

South Carolina

Mount Pleasant, transfer to the train and be on the Isle of Palms before lunch.

This lush oasis has not always been such a tranquil retreat. Great crowds of people enjoyed swimming and picnicking; there were fireworks, a ferris wheel and dancing to the music of nationally known orchestras. At low tide motorcycles ripped up and down the beach competing for prizes. It all came to a screeching halt after World War I when the financially troubled railroad ceased operation, and the island was again isolated. The few residents had to walk across the railway bridge to Sullivans Island or travel by boat. Bridges built in the 1920s opened the island to development, and a permanent community became incorporated in the 1950s.

A few years ago the Isle of Palms Beach and Racquet Club began implementing their well-thought-out plan for a private residential community with the focus on leisure time activities. Great care was taken to preserve the natural beauty of the area. Groves of live oaks were left untouched, as was much of the dense jungle-type growth which gave the island its name.

Bike and nature trails wind past tennis courts and fishing lagoons. Those who prefer freshwater will find at their disposal a junior olympic swimming pool; those who like the ocean have the run of two and a half miles of beach. The children's program includes historic tours, ice cream churning and kite flying along with the standard favorites—swimming, crabbing and sand castle construction. Charter boats for ocean fishing are available at nearby Shem Creek.

Sullivans Island

How to get there
Car: From Charleston, take U.S. 17N (business) to Mt. Pleasant, turn right on S.C. 703.

Activities and accommodations
Swimming
Fishing
Sightseeing: Fort Moultrie and Charleston.
Accommodations: Rental cottages.

For more information
Visitor Information Center, 85 Calhoun St., Charleston, S.C. (803-722-8338).

Sullivans Island, eight miles east of Charleston, South Carolina, is a favorite summer retreat for Charlestonians. This quiet residential island community is comprised of a mixture of rental cottages,

Sea Islands of the South

summer homes and elegant permanent residences. The draw, of course, is the beach experience—pelicans flying in squadrons just above the wave troughs, sun toasting teenagers golden brown, and moon shimmering on the restless Atlantic.

For many, the beach is enough. Others prize the additional bonus, the proximity of the charming, historic city of Charleston with its fine restaurants, big-city conveniences and impressive selection of sightseeing attractions.

Those who look to Charleston for all their sightseeing will miss Sullivans Island's own **Fort Moultrie**. This well-maintained national monument has a series of fascinating stories to tell and is well worth an hour or two. After taking S.C. 703 to Sullivans Island, turn right on Middle Street. The fort is two miles from the intersection. Moultrie has been restored so each of the five areas represents a different period in the history of seacoast defense in the United States—1809-1860, Civil War, the 1870s, 1898-1939 and World War II.

Here in 1776 Colonel William Moultrie and about 400 brave South Carolinians held off a fleet of nine warships while a small force of colonists blocked an infantry attack at the island's north end. The original fort, a tiny 500-foot square palmetto log and sand fortification, was only half-completed when the British attacked. The battle proved to be the first major Revolutionary War victory for the Americans. The news, which reached Philadelphia a few days after the Declaration of Independence was adopted, elicited a statement of gratitude from the Continental Congress. The South Carolina legislature named the fort in honor of its commander.

The tradition of seacoast defense on Sullivans Island which started with that 1776 victory did not end until 1947. The fort as it stands today was completed in 1809 and is the third fort on the site. During World War II Moultrie became a reinforced concrete fortification.

Wars did not provide the only drama. The fort was badly damaged by hurricanes in 1783 and 1804. Edgar Allan Poe used the locale for his well-known short story *The Golden Bug*, which was written while the famous author was stationed at Moultrie. In 1838 a group of Seminole Indians being moved west was imprisoned here. Osceola, a great Indian leader who died within a month of his imprisonment, is buried just outside the main gate.

After Pearl Harbor the fort was updated. New 90-mm anti-aircraft guns were added, but within a few years technological advances in warfare rendered the fort obsolete. Two centuries of coastal defense on Sullivans Island came to a close with the 1947 deactivation of Fort Moultrie.

Pick up a free brochure at the visitor center which explains the fort's history and browse among the interesting exhibits. (Daily. Summer 9-6, Winter 9-5, free.)

South Carolina

Photo courtesy of Kiawah Island

Folly Island

How to get there
Car: Take 17 south from Charleston and turn left on Rte. 171. Follow Folly Road to Folly Island.

Activities and accommodations
Swimming
Fishing and crabbing.
Sightseeing: Charleston!
Accommodations: Motels, rental cottages.

For more information
Visitor Information Center, 85 Calhoun St., Charleston, S.C. (803-722-8338), Holiday Inn, 108 Ashley Ave., Folly Island, S.C. 29439 (803-588-2191), Sellers' Shelters, 104 West Ashley Ave., Folly Island, S.C. 29439. (803-588-2269).

Folly Beach has been a family retreat for Charlestonians for a long time. A public beach, a community of stilted cottages and a few motels cluster here on the Atlantic where the attractions are the sun, sand and surf. Shelling is good as is the swimming, crabbing

Sea Islands of the South

and surf fishing. The marsh side of the island is a haven for shore birds.

Besides the usual island activities, Folly offers quick access (20 minutes) to Charleston, one of the oldest cities in the United States featuring history, culture, fine restaurants, excellent golf, elegant plantation homes and some of America's most spectacular gardens. When you've had enough sun, just brush the sand from between your toes and head for town!

Kiawah Island

How to get there
Car: 20 miles southwest of Charleston, S.C., via State 7 and U.S. 17 to Main and Bohicket Roads that cross Johns Island.
Plane: Business, charter or private planes use Johns Island Airport, 15 minutes away.

Activities and accommodations
Golf
Tennis
Biking
Swimming: Pool and ocean.
Fishing
Boating: Charter boats available nearby.
Children's activities: Outstanding variety of activities for tots and teens. Well supervised.
Sightseeing: Jeep safari to Vanderhorst Plantation home. Water tour of Canvasback Lake. Charleston!
Accommodations: Rooms in Kiawah Island Inn and villas.

For more information
Kiawah Island Resort, P.O. Box 12910, Kiawah Island, S.C. 29410 (803-559-5571).

Kiawah Island, 21 miles south of Charleston, South Carolina, may be reached via Interstates 95 and 26. Here you'll find first class resort facilities in a setting of incredible natural beauty. Those who want a day on Kiawah's beach without the price tag of an overnight stay, should plan a picnic at Beachwalker Park.

This slender 10,000-acre barrier island was named for the Indians who hunted and fished here until the white men appeared on the scene. After that the Kiawahs didn't last long. Those who weren't shot or enslaved succumbed to smallpox and measles imported by the Europeans. In the 1690s pirates stashed their loot on this densely-wooded isle, and later Englishmen planted cash crops in the island's rich soil. The earliest harvests were of indigo, but the

South Carolina

Roscoe Tanner holds a teaching clinic at Kiawah

Revolution soon cut off the English market for the valuable blue dye so planters turned to Sea Island cotton. From 1772 to 1953, the island was privately owned by Arnoldus Vanderhorst, the seventh governor of the state, and his descendants. They were referred to as the "Kings of Kiawah" by the local populace in Charleston.

After the Civil War, Charlestonians enjoyed all-day excursions to Kiawah by passenger packet. These beach picnics were immortalized in the folk opera *Porgy and Bess* written by Dubose Heyward while he lived in Charleston. In the early '50s a lumberman purchased Kiawah for $125,000. His heirs sold it to the Kuwait Investment Coroporation in 1974 for $17.3 million.

The oil-rich investors decided to do exhaustive research before turning the first shovel of earth in order to best preserve the island's natural riches. The 16-month, million-dollar conservation study utilizing scientists from 13 disciplines turned out to be the largest ever conducted by a privately-held island. There was much to investigate. Forests of live oak, magnolia, pine and palmetto provided habitat for white-tailed deer, river otter, bobcat, raccoon, opossum and a herd of wild horses. Over 150 species including the endangered brown pelican and the Atlantic loggerhead sea turtle, thrive on the island. Steps were taken to protect wildlife from human interference and certain areas were set aside to remain forever wild.

Sea Islands of the South

One of the most rewarding programs to come out of the study was the sea turtle patrol. Marine biology graduates and students monitor adult turtles when they come ashore during summer nights to nest. Interested guests are welcome to watch the collecting of eggs. The eggs, carefully placed in a sand-filled bucket, are covered with a sheet to simplify observation and a layer of sand for insulation. The buckets are placed in a hatchery where they are watered every few days. After 50 days, the nests are checked daily so emerging hatchlings can be escorted safely to the ocean. In its first five years of operation the turtle patrol returned more than 11,000 tiny turtles to the sea.

Since lights are a major deterrent to turtles ready to nest, special non-glare lighting with indirect focus was designed for the Kiawah Inn. Also, guests are reminded to keep their drapes closed during nesting season. These measures have proved to be very successful; several turtles came ashore directly in front of the inn and sailboats had to be quickly moved out of their way.

Another discovery that came from the scientific study was that Kiawah is growing rather than suffering from erosion problems that plague most barrier islands. When Charleston harbor jetties were constructed 20 miles to the north in 1896, sediments that used to land on nearby islands were diverted and deposited in deeper waters. This meant greatly accelerated erosion for those islands but

Gary Player designed this 18-hole course at Kiawah Island complete with lagoons, lakes and saltwater marshes. (Photo courtesy of Kiawah Island)

South Carolina

The 18th century Vanderhorst Mansion (Photo courtesy of Kiawah Island)

added eight to ten feet a year to Kiawah's eastern shore.

Two tours, the **Back Island Safari** and the **Water Tour**, acquaint visitors with the rich wilderness of this subtropical island. Some guests get to drive a jeep in the caravan that bumps down old logging trails, around alligator ponds and along the beach. The guide's commentary over CB radio covers island history (including a ghost story) and ecology. The two-hour trip includes a visit to the historic Vanderhorst Plantation home, circa 1772. This sturdy mansion made of cypress weatherboarding has withstood an earthquake, two hurricanes and occupation by both Union and Confederate troops. Those interested in closeup views of alligators, large wading birds, jumping fish and turtles should take the water tour which explores the unique scenery of Canvasback Lake. (For reservations write Kiawah Island, P.O. Box A101, Kiawah Island, S.C. 29455 or call 803-559-9011.)

Kiawah's facilities for golf and tennis are superb. Guests may also use the ocean-front pool, men's and women's spas with whirlpool, sauna, showers and lockers. Winding nature trails beckon bikers and hikers and a jogging course challenges fitness buffs with exercise stations along the way. Programs for both children and teens are well supervised and offer a potpourri of activities from disco dancing to crabbing. You can charter a boat for deep sea fishing, join an astronomy lecture or chat with your island

Sea Islands of the South

neighbors at an oyster roast. In short, there's no shortage of things to do!

Casual fare is served at outdoor tables in front of the Ice House Cafe, or you may want to lunch on the Jasmine Porch where slow-moving fans and hanging plants create a relaxing mood. Gourmet seafood is served in the evening at the elegant Charleston Gallery. All restaurants are within a three-minute stroll of your room.

This is one resort where you really don't need a car since everything is so convenient. Bikes can be rented and the tours and transportation office can arrange limousine, bus or rental car transfers for short trips to Charleston and other off-island areas.

Those who prefer to experience Kiawah for the price of a parking fee will enjoy **Beachwalker Park**. Open daily from 9:30 to 6:30 from Memorial Day to Labor Day, the park offers rest rooms, dressing areas, outdoor showers, trails, picnic tables and the main attraction—the beach. Animals and surfboards should be left at home.

Seabrook Island

How to get there
Car: 23 miles south of Charleston, S.C., via U.S. 17 South and S.C. 700.
Plane: Private. John's Island Airport (rental cars available).

Activities and accommodations
Golf
Tennis
Biking: Rentals available.
Swimming
Fishing and crabbing
Horseback riding
Rental boats
Children's activities: Many varied activities, well-supervised.
Sightseeing: Charleston!
Accommodations: Cottages and villas.

For more information
Seabrook Island Company, P.O. Box 32099, Charleston, S.C. 29407 (800-845-5531).

Seabrook Island, 23 miles from Charleston, is just off Johns Island and across the creek from Kiawah. This carefully-planned, controlled-access refuge (for people as well as wildlife) is home of the Seabrook Island Resort. To its natural blessings—three miles of splendid beach and virgin subtropical jungle—man has added every

South Carolina

possible recreational facility.

The Kiwatos Indians who hunted here stayed around long enough to consume huge quantities of oysters, leaving their shells in great mounds. When Lt. Col. Robert Sanford claimed the island in 1666 for the English crown, he also "took formal possession by turf and twig of the whole country from the latitude of 36 N to 29 SW to the South Seas." When you add the rest of the country to this 2000-acre island, it was quite a chunk of real estate! During colonial times battles were fought between France and Spain for control of the land and, just as on other sea islands, pirates sequestered their stolen goods here.

With profits from his Sea Island cotton, William Seabrook bought the island from another wealthy planter. His plantation was on nearby Edisto Island so at first he used it primarily as a hunting preserve. Around 1750 he built a fine home which still stands across the causeway from the island. His descendants sold the property in 1863 and it then passed through several hands before being purchased by the Seabrook Island Resort in 1971.

Unlike most other sea islands, Seabrook was never timbered. Dense forests of stately palms and live oaks draped with Spanish moss shelter all kinds of wildlife including whitetail deer, rabbits, squirrels, raccoons, opossums and bobcats. Alligators laze beside quiet lagoons and the marshes are alive with herons, egrets and marsh wrens. Most common shore birds are gulls, terns and scoters; woodpeckers, cardinals, warblers, vireos and many other species can be found inland. The best performers are the dive-bombing brown pelicans. Deveaux Banks, just offshore, is a sanctuary for this rare and endangered species.

The theme at Seabrook is preservation, and construction blends with the natural environment. Roads meander around ancient oaks, the marshes are inviolate and the natural contours of the land have not been altered. Retirement homes, rental villas and vacation hideaways are built in accordance with a strict code so as not to detract from the island atmosphere.

What is there to do? The list is staggering! Newcomers, naturally drawn to the ocean, can choose from a variety of activities. Shellers gather whelks, cockles, angel wings and sand dollars by the dozens. The hard-packed sand makes biking popular. Rent a sailboat or take a dip in the gentle surf. Swimmers who prefer fresh water have their choice of three oceanside pools. Those who just want to relax in the sun have the view to end all views—shrimp boats on the horizon, squadrons of pelicans and porpoises frolicking less than 20 yards offshore.

Activity for fishermen of every temperament includes surf casting for channel bass, bluefish, mullet, flounder and speckled trout, freshwater angling for bass and bream and deep sea fishing for Gulf Stream monsters. Tidal creeks abound with blue crabs, shrimp, oysters and clams. Fishing tackle, crabbing gear and bait are available at the tennis shop and you can charter a boat at nearby Stono Marina on Johns Island. (803-559-9011 for reservations.)

Seabrook's 18-hole golf course is known for its view as well as its natural hazards—dunes, marshes, creeks and unpredictable sea breezes. You'll find the tennis center quite complete with seven courts, ball machine, racquet stringing services and a tennis shop. Volleyball, horseshoe and shuffleboard equipment are available. Teens enjoy Ping-Pong and other games in the recreation room and a full schedule of supervised children's activities offers everything from bike safaris to sand castle construction.

If you're interested in coastal ecology, join one of the guided beach walks (803-559-5511, extension 233, for schedule). Or explore the woods and marshes on horseback as you follow the Swamp Fox nature trails which were named for Revolutionary War hero General Francis Marion. Guests may sign up for Tally Ho jump rides, beach rides, group or individual instruction. (Riding information and reservations, 803-559-5511)

Photo courtest of Kiawah Island

Deveaux Banks
Audubon Society Sanctuary for Brown Pelican

Visitation: None

A crescent-shaped sand bar visible from Seabrook's Beach Club

South Carolina

on Seabrook Island, Deveaux Banks is an official Audubon Society sanctuary for the endangered brown pelicans. These impressive creatures, four and a half feet long with a seven-foot wingspan, are our largest sea birds. Their fate was severely threatened by the use of DDT, but they have made a noticeable comeback since the pesticide was banned. Time spent watching pelicans divebomb for fish or fly single file just above the wave crests is well used. Their kamikaze fishing style may seem like a difficult way to corral breakfast, but it gets results. They dive headlong into the ocean to spear menhaden, mullet or herring with their pointed beaks, or they may net the fish in their stretchy pouches which can hold eight quarts of water. After a quick return to the surface, insured by special inflatable air sacs under the skin, the pelican swallow their catch head first to avoid sharp scales or spines.

Deveaux Banks is one of the northern-most rookeries for the brown pelican and one of the two largest breeding grounds in South Carolina. The Audubon Society recently dedicated Deveaux Banks as the Alexander Sprunt Memorial Sanctuary.

Edisto Island

How to get there
50 miles southeast of Charleston on S.C. 174.

Activities and accommodations
Golf
Tennis
Biking
Swimming
Fishing and crabbing
Boating: Rentals available.
Sightseeing: Charleston!
Camping: Cottages and campsites at Edisto Beach State Park.
Accommodations: Rental cottages, lodge cabins (Oristo Resort).

For more information
Edisto Beach State Park, Route 1, Box 40, Edisto Island, S.C. 29438, 803-869-2156. Oristo Resort, P.O. Box 27, Edisto Beach, S.C. 29438, 803-869-2561.

Edisto Island offers a settled, easy-going atmosphere reminiscent of mint juleps on shaded porches. A wide range of facilities in a setting of rare subtropical beauty awaits the visitor. You may rent a stilted beach cottage, camp near the surf at Edisto Beach State Park or stow your golf clubs in your own private lodge cabin at Oristo and enjoy all the amenities of a full-fledged resort.

Sea Islands of the South

This lush island, with stately old plantation homes and jungles of palmetto, myrtle, yucca and jack vine, is bounded by two branches of the Edisto River, the Dahoo River and the sea. Beachcombers who are allergic to crowds revel in the shell-packed stretches of sand, and fishermen face agonizing decisions—whether to surf cast, go offshore or try the lagoon.

The peaceful, agrarian Edisto Indian tribe once pitched their tents on this bountiful island. They found the sea crowded with oysters, crabs, clams and fish, and the forests teeming with game. Although Robert Sanford took possession of Edisto for King Charles II in 1666, it wasn't until 1674 that the Earl of Shaftesbury officially purchased the island from the Indians for a piece of cloth, some hatchets, beads and other goods. The Spaniards called the island Oristo when they raided it in 1686, but the English named it for the Indians when they reclaimed the land for the crown.

The first permanent settlers grew indigo after finding local conditions unsuitable for rice. When the profitable English market for the blue dye dried up with the American Revolution, plantation owners switched to Sea Island cotton. Edisto cotton never did get to market. Textile mills in France contracted for the entire crop before the seeds were put in the ground. The French needed the world's finest cotton to produce their famous high quality cloth. Edisto planters carefully perfected and took great pride in their own strains of seed cotton; they could be counted on for an outstanding crop.

For their trouble plantation owners were rewarded with enormous sums of money, enough to live in considerable grandeur. They built elegant homes, kept town houses in Charleston and furnished them with imported mahogany and rosewood furniture. Their formal gardens blossomed with rare trees and exotic plants, they entertained on an incredibly lavish scale and their children were sent to Europe to be educated.

When Port Royal was taken over by Union Forces in November 1861, the governor ordered all women, children and disabled men to leave the island. With the workers already at war, the fields quickly became overgrown. Edisto's once-proud mansions, monuments to the fabulous era of Sea Island cotton, were vandalized by Yankee soldiers. Some of these homes still stand, a few in ramshackle condition, others lovingly restored.

Today the island's riches are its white sand beach, vast tidal marshes veined with meandering salt creeks and inlets, a thriving bird and wildlife population and a quiet community of residents who like Edisto just the way it is.

Edisto Beach State Park, 50 miles south of Charleston on Rte. 174, is a fine spot to make an initial acquaintance with the island. Its two-and-a-quarter miles of beach is considered one of the best in the state for swimming and shell collecting. Sometimes

South Carolina

beachcombers find fossilized fragments of bison, mastodons, giant armadillo and three-toed horses. Campsites are set in attractive groves of cedar, crepe myrtle and some of the tallest palmettos in the state.

The South Carolina Institute of Archaeology and Anthropology has recorded ten prehistoric sites on Edisto Island. A four-mile self-guiding trail in the state park leads across a marsh to one of these sites, an Indian shell mound. No one knows if these mounds were used for ceremonies, burials or just refuse.

The more visitors learn about marshlands, the more they appreciate their importance and their beauty. Once thought of as wasteland, the salt marsh is actually vitally important to many forms of life including our own. Shrimp come to the shallow, nutrient-rich waters to mature though they go to sea to spawn. Fiddler crabs, on the other hand, spawn in the marshes, then go to sea. Some animals spend their entire lives in the tidal waters, while others make brief visits in search of food. Most marine life is directly or indirectly dependent on the marsh.

Crabbing, clamming and fishing are favorite pastimes. Offshore catches include Spanish mackerel, dolphin, yellowtail, red snapper, blue marlin and grouper. A great variety of birds may be seen in any season along with deer, squirrel, raccoon, oppossum, rabbit, fox and wild turkey. During late September and early October the southern end of the island is literally covered with butterflies as thousands of Monarchs migrate south.

Oristo, a 300-acre family resort, is set in the tangled subtropical forests on the island's southern tip. Its temptations are an outstanding golf course studded with white sand bunkers and shaded by

Sea Islands of the South

great live oaks, tennis courts, jogging trails, rental bikes, a swimming pool and tastefully designed cottages that blend with the landscape. Umbrellas, floats and sailboats may be rented at the Oristo beach cabana which has showers, dressing rooms, picnic tables and a large deck for sunning.

The island is perfectly situated if your definition of convenience is crab succulent with net-your-own freshness, but not if you need a supermarket down the street. Visitors to Edisto Island know how to entertain themselves. They walk the beach by moonlight or make a project of listening to the night sounds. Those who hanker for an occasional night on the town head for the lights and bustle of Charleston. But real island lovers are always glad when they can wend their way back home.

Hunting Island

How to get there
16 miles southeast of Beaufort, S.C. on U.S. 21.

Activities and accommodations
Carpet Golf
Swimming
Fishing and crabbing
Boating: Rentals available.
Children's activities: Supervised recreational activities.
Sightseeing: Historic Beaufort.
Camping: Campsites and cabins in Hunting Island State Park.

For more information
Beaufort Chamber of Commerce, Box 910, Beaufort, S.C. 29902. (803-524-3163) Applications for cabins should be addressed to: Director of State Parks, P.O. Box 1358, Columbia, S.C. 29202. From Labor Day until May 31 reservations for cabins may be obtained by writing directly to Hunting Island State Park, Frogmore, S.C. 29920.

Hunting Island State Park is the main attraction of lush subtropical Hunting Island. This 5000-acre park offers three miles of wide sandy beaches, oceanfront camping, vacation cabins, public boat launching facilities, artificial fishing reefs, hiking and nature trails, carpet golf and a view of it all from an old lighthouse.

This island was once used almost exclusively for hunting, first by the Indians and later by the colonists, which accounts for its name. (Hunting is no longer allowed.) Like other sea islands, this one is constantly changing shape according to the whim of wind, tide and surf. The foundation of the original 1859 brick lighthouse which

South Carolina

was destroyed by beach erosion can be seen from the top of the present 136-foot **Hunting Island Lighthouse**. (Open daily from 12 noon until dusk). The U.S. Coast Guard replaced the lighthouse in 1875 constructing it of cast iron plate so, if necessary, it could be moved. By 1889 the site which had once been a quarter mile from shoreline was again threatened by erosion and the new light erected in its present location. Today's lighthouse is a point of interest for view-minded visitors; its use as a navigational aid was discontinued in the 1930s.

You can better appreciate the complexities of coastal ecology if you take the one-mile **Lighthouse Trail** which features a 1000-foot elevated boardwalk over marshes and interior lagoons. The island is a refuge for deer, raccoon, migratory waterfowl and all sorts of small game. Hunting's beach is a nesting site for the endangered Atlantic loggerhead sea turtle which comes ashore during summer nights to lay her eggs.

Lagoon fishing is productive throughout the year, and surf fishing is especially popular in early spring and late fall when there is a good run of whiting, trout, bass and drum.

Although the beach has suffered from extensive erosion, it is still one of the finest on the coast. Designated areas protect swimmers from surfers and surf fishermen. Sea shell collectors find a bonanza of starfish, conch shells, sea pens, cockle shells, oysters and sand dollars.

A hundred trailer and tent sites have water and electrical hookups, picnic tables and grills. Conveniences include rest rooms with hot showers, a general store and refreshment stand. Campers are welcome on a first come, first served basis and are limited to a week's stay.

Due to heavy demand for cottages from June 1 to Labor Day, a drawing of applications is held just after March 1 each year and those selected may reserve cabins for a week. Applications should be addressed to: Director of State Parks, P.O. Box 1358, Columbia, S.C. 29202. From Labor Day until May 31, reservations for cabins may be obtained by writing directly to Hunting Island State Park, Frogmore, S.C. 29920.

Fripp Island

How to get there
Car: 19 miles southeast of Beaufort, S.C. Follow U.S. 21 from Beaufort through Hunting Island State Park to the Fripp Island Bridge.

Plane: Private aircraft may utilize the Beaufort County airstrip.

Sea Islands of the South

Activities and accommodations
Golf
Tennis
Biking
Swimming: Pools and ocean.
Fishing
Boating: Rental sailboats and charter boats for deep sea fishing.
Children's Activities: Extensive program, well supervised.
Sightseeing: Historic Beaufort.
Accommodations: Villas and tree houses.

For more information
Fripp Island Resort, Fripp Island, S.C. 29920. East of Mississippi River, call toll free 1-800-845-4100. In South Carolina and west of the Mississippi, call 803-838-2411.

Fripp Island, a 3000-acre wooded barrier island, is a fine place to shape up or slow down. Outstanding recreational facilities lure visitors who crave action, and tranquil subtropical beauty appeals to those who prefer unwinding at a leisurely pace.

The Yemassee Indians didn't need a full-color brochure to entice them to come hunt and fish on the island. Nor did the pirates who hid their plunder under the palmettos before sailing off into the sunset. During colonial times Captain Johannes Fripp, who risked his life protecting the coastal waterways from the Spanish, was awarded the island in return for his efforts. Today Fripp Island is a private, fully serviced residential community where southern hospitality is a way of life. Guests have a choice of accommodations in a villa, tree house or Fripp Island Inn room.

This ecologically-sensitive resort was carefully developed to enhance its natural gifts. These assets include three and a half miles of uncrowded beach, three large tidal marshes and tropical forests sheltering a wide variety of wildlife. Visitors receive an environmental bulletin with island regulations concerning the protection of flora and fauna.

The endangered Atlantic loggerhead sea turtle has become a community project among Fripp Island residents. The mother turtle comes ashore on summer nights to lay her eggs above the high tide line. Lookouts are posted for the nests and once discovered, the eggs are protected from predators. When the half dollar-size hatchlings emerge, residents escort them to the sea to keep hungry birds from picking them off.

Be sure to get an activities schedule and weekly calendar of events when you arrive. From oyster roasts to family fun runs, something is always going on. The Spring Festival, held every year in April, is a "low country" fair complete with sidewalk art show,

South Carolina

antique musical instrument display, puppet show, square dancing and wine tasting. Try your hand at such skills as sand sculpture, apple bobbing and nail driving! The annual Bass Fishing Tournament is in October and an art exhibit is held each spring. Every March, Ralph Smith, nationally known painter, holds a six-day watercolor workshop. In May you may watch the island's Laser sailing regatta which has been selected as one of the five qualifying races for the Laser Worlds.

As much fun as these events are, every day on the island offers enough opportunities for pleasure to be labeled "special." You might begin by taking a pulse-quickening jog along the parcourse, spend the day at golf, tennis, swimming or deep sea fishing, relax watching the sun disappear on the horizon and after a fresh seafood dinner, see a theatrical show by a professional touring group. Not ready for bed? Take a stroll down the beach. You might come upon some deer in the moonlight.

The parcourse jogging trail, consisting of 18 stations with apparatus designed for both children and adults, is one of the best on the coast. Signs clearly describe and illustrate each specified exercise. A recommended "par" or number of repetitions is suggested on each sign. Starting with warm-ups and progressing to more demanding chin-ups and vault bar work, the participant can gradually increase the number of repetitions at each station and the speed at which he or she runs the course. Of course, you might just enjoy walking the scenic one-and-a-half-mile trail which winds through swaying sawgrass and under ancient live oaks and tropical palmettos.

Tennis buffs will appreciate a choice of hard or soft courts (14 in all), a fully stocked pro-shop, ball machines and a video tape machine. You may sign up for lessons, clinics and weekly tournaments. A player rating system ensures you a partner of near-equal ability.

The 18-hole championship golf course has been described as one of the most picturesque on the eastern seaboard as well as one of the most challenging. Four holes border the ocean including the infamous 18th which is 511 yards.

The Fripp Island Marina offers dockage, charter boats for deep sea fishing and a well-stocked tackle shop. Rental sailboats and bikes are available and swimmers have a choice of three pools. Children may join a program of well-supervised fun including nature walks, biking, shelling, arts and crafts.

Hilton Head

How to get there
Car: 31 miles from Savannah, Ga., and 90 miles from Charleston,

Sea Islands of the South

S.C. 40 miles east of I-95. Take interstate expressways to the Savannah or Charleston areas, then U.S. 17 and follow signs.

Plane: Coastal Plains Commuter planes provide daily service from island airport to Savannah and Charleston airports making connections with major airlines serving those cities. The island has a 3700-foot paved runway airport which accommodates most non-jet private aircraft.

Boat: On Intracoastal Waterway. Excellent facilities.

Activities and accommodations
Golf: Outstanding.
Tennis: Outstanding.
Biking: Rentals available.
Swimming: Pools and ocean.
Fishing and crabbing: Every type available.
Horseback riding: Three stables. Many miles of trails.
Boating: Excellent facilities. Charter and rental craft available.
Children's activities: Various resorts have supervised summer youth programs.
Sightseeing: Historic sites, cruises.
Camping: Motorhome resort.
Accommodations: Motels, hotels, lodges, rental condominiums, villas, houses and apartments.

For more information
Hilton Head Island Chamber of Commerce, P.O. Box 5647, Hilton Head Island, S.C. 29928 (803-785- 3673).

South Carolina

Hilton Head, the largest oceanfront island between New York and Florida, is just off the southeastern tip of South Carolina approximately 30 miles north of Savannah, Georgia. This twelve-by-five-mile island, home to one of the finest resort communities in North America, has become the best known of the Sea Islands.

Obviously Hilton Head is not an undiscovered wonder. The island golf and tennis tournaments have received a great deal of television coverage and its developers have mounted an extensive publicity campaign. About 10,000 permanent residents are inundated by a huge wave of visitors during the peak summer season. If you're looking for a quiet little out-of-the-way place, you're about twenty years too late. This is not to say the Hilton Head experience won't be a quality one. In fact, quality has obviously been the theme of its development. You won't find any garish signs and honky-tonk shopping centers here. McDonald's golden arches are tastefully disguised, the 7-11 store sedately announced its presence with roman numerals and unobstrusive low country architecture prevails.

Your memories of Hilton Head will be personal ones. They may include beachcombing for sharks' teeth, watching a Sunday afternoon rugby-polo match or biking through a cool forest of moss-draped live oaks. You'll need a lifetime to take full advantage of all the island's offerings, a few of which include 12 miles of white sand beach, 11 championship golf courses, tennis, biking, hiking, sightseeing, horseback riding and boating.

Despite the smorgasbord of resort facilities, thousands of acres have been permanently set aside as wilderness and wildlife sanctuaries. When the developers preserved large tracts of land as inviolate forest and marshland, they saved the habitat of deer, raccoons, opossums, gray foxes, squirrels, bobcats, wild turkey and alligators.

Two of the remaining 13 waterfowl rookeries in this coastal area are found on Hilton Head Plantation. **Cypress Swamp Rookery,** nesting area for the endangered osprey, is closed but can be viewed from a 60-foot tower. **Whooping Crane Pond** is home to hundreds of white ibis, egrets and a variety of wading birds. A leisurely hour's walk around the waterfowl pond loop at **Sea Pines Forest Preserve** provides an excellent closeup view of ducks and herons. Bird watchers can have a field day on this island with more than 260 species on the wing.

Be sure to explore **Heritage Farm** on the southwest boundary of Sea Pines Forest Preserve. Here you can learn about organic farming and the old plantation-style agriculture. A particularly fascinating "endangered" farm institution is the mule-powered mill where sugar cane is ground and boiled into syrup.

The waters surrounding Hilton Head are brimming with life. Shrimp boats and oyster bateaux bring sea-fresh shellfish to market

Sea Islands of the South

and sport fishermen hook into one challenge after another. Bordered by the last major unpolluted marine estuary on the East Coast, the island is famous for both salt and freshwater angling. Surf fishermen with 12 miles of beach at their disposal cast for everything from winter trout and flounder to huge 100-pound drumfish. The most productive spot for the channel bass run in early spring and late fall is the island's south end where bottom-stirring currents create a natural feeding ground. Ponds and lagoons are stocked with largemouth bass and bream. Charter boats can reach the gulf stream in a few hours if you're after sailfish or marlin, but the blackfish banks and mackerel grounds are closer and also promise plenty of action.

High powered boats and elaborate fishing gear are fine, but crabbers have been known to have an awful lot of fun with a line, a dipnet, a chicken neck and a little patience. And fresh blue crab is hard to beat as a palate pleaser.

The **Jean-Michel Cousteau Institute,** established here in 1977, is dedicated to the well-being of mankind. This public non-profit foundation hopes to create a strong desire to preserve our fragile ocean environment through a process of education. Expeditions, seminars, conferences and a series of interesting oceanology workshops have started to work toward this end. (785-3333).

Beachcombers have learned the best time to hunt shells is at low tide when nearly 600 feet of beach is exposed. The northeast end of the island yields an especially good variety including the standard starfish and sand dollars.

The collection of 11 championship golf courses is as fine as you'll find any place in the world. Since 1969 when Arnold Palmer won the first Heritage Golf Classic, island golf has drawn a national audience. Twice a year, Hilton Head hosts the world's best golfers in the Heritage and Women's International tournaments.

Other attractions are a six to eight month swimming season, 12 miles of paths and rentals for the bike crowd and three well equipped stables. The Sea Pines Hunter Classic Horse Show is a gala social gathering.

The island has a rich historic background going back as far as 10,000 years when wandering hunters stalked game in the forests. Archaeologists have discovered remnants of four different Indian tribal cultures on Hilton Head. Traces of the Cusabo Indians indicate they were a peaceful people who farmed vegetables, fished and hunted about 4000 years ago. Their simple life was rudely interrupted by the Spanish and French explorers who fought for control of the harbor in the 16th century.

The island's name changed a number of times from Punta de Santa Elena to Isla de los Osos (Island of the Bears) and Ile de la Riviere Grande (Island of the Big River). Then Sir William Hilton happened on the scene. The English sea captain sailed into Port

South Carolina

Royal Sound in 1663 in search of fertile farmland. His description was euphoric: "It is most pines, tall and good. The ayr is clean and sweet, the land passing pleasant...the goodliest, best and fruitfullest isle ever was seen." The bluff he stood on to view the harbor was known as a head and eventually the whole island became Hilton Head.

In the 1700s plantations devoted to growing indigo, rice and sugar flourished. When the English market for indigo disappeared completely during the American Revolution, the planters turned to Sea Island cotton. In 1825 this cotton, highly prized for its silky texture, sold for $54 a pound compared to $12 for regular cotton. Plantation owners became wealthy but their luxurious lifestyle was to come to an abrupt end with the Civil War.

In 1860 South Carolina seceded from the Union. Fort Walker, constructed primarily by slave labor donated by plantation owners, was to defend the coast. However, Union forces took over the fort when they defeated the Confederates in the decisive Battle of Port Royal on November 7, 1861. The North used the island as a base to blockade Southern ports, particularly Savannah and Charleston. A wartime community of 25,000 people sprang up which supported two newspapers, hotels, shops, a theatre and a hospital.

After the war Hilton Head was taken over by a small freed slave population who farmed, fished and hunted here. Some large tracts of land were later bought by hunting clubs for the "ayr" was still clean and sweet and the land densely forested. In 1950 two men interested in buying property for timberland formed the first development company, the Hilton Head Company.

A bridge was built to the mainland in 1956 and that same year Sea Pines Company was formed by Joseph Fraser and his sons. They believed land could be developed without being destroyed, that natural beauty should be preserved and that a community was best served by a carefully designed overall plan. Golf courses, tennis courts, marinas and swimming pools were built first, homes and businesses later. It was a revolutionary idea whose time had come. Property values began to soar as demand ran high. Beachfront lots which sold for $300 in 1960 went for $150,000 in 1978. Once unknown, Hilton Head had gained an international reputation as a premier resort and residential community in a brief 20-year span.

Sightseers may want to tour first with an experienced guide before striking out on their own. **Travel Venture Tours** (803-785-5237) offers a two-hour guided introduction to Hilton Head, and Ann Parker Tours (785-7273) highlights recreational and residential areas as well as historical sights.

You may prefer a do-it-yourself tour. Don't be confused by the term plantation. The entire island was divided into these large agricultural land holdings in the lush period before the Civil War. Today these "plantations" are actually developments, some residential,

Sea Islands of the South

some primarily resort.

Several historic sites are located inside Port Royal Plantation. After going through the gate, take the first road to the left and follow it to the historical marker. You'll see the earthen mounds and decaying fortifications of **Fort Walker** which was overcome by Union forces in 1861. Walker was the principal base of the northern naval blockade, perhaps the greatest single factor in the defeat of the Confederacy. You'll also see the concrete foundation and a few recoil tracts nearby of one of the first steam cannons. This particular one was test fired once; the resulting explosion started a forest fire and that was the last experimentation with this type artillery.

The **Baynard Ruins** are located in Sea Pines Plantation near the intersection of Baynard Park Road and Plantation Drive. William Baynard won 1000 acres called Braddocks' Point in a poker game and built a beautiful plantation home here in 1830. The house was accidentally burned after the Union takeover in 1861. But the remains of its massive tabby walls and some of the outbuildings give some idea of the size and scope of the typical 18th-century low country plantation.

Baynard Mausoleum is the largest, intact antebellum structure on Hilton Head. It's located at the intersection of Route 278 and Matthew's Road in the Zion Cemetery where many of the island's

South Carolina

early families, including several Revolutionary War veterans, are buried.

Fort Mitchel is in Hilton Head Plantation adjacent to the Old Fort Pub. Union forces constructed this fort on Skull Creek to protect Port Royal Sound. Interpretive signs along a boardwalk explain the moat, cannon and bomb-proof shelters.

To better appreciate Hilton Head's unique ecosystem of ocean, marsh and woods, hop aboard the *Vagabond* for a cruise of Calibogue Sound. You'll get a good look at nearby Daufuskie Island and probably see osprey, wood storks and dolphins. (Call 803-785-4684 for information about afternoon and sunset cruises.) Another fun boating excursion is the Daufuskie Island lunch tour which features deviled crab and other low country delicacies. (785-5236 or 785-5237.) Cruising tours depart from Harbour Town.

Accommodations on Hilton Head include lodges, motor inns and hotels as well as rental condominiums, villas, houses and apartments. Recreational vehicles can choose from more than 400 sites in the country's first motorhome resort.

Daufuskie Island

How to get there
Boat: From Harbour Town, Hilton Head, by cruise boat. (See below.)

Activities and accommodations
No overnight accommodations.

For more information
Hilton Head Island Chamber of Commerce, P.O. Box 5647, Hilton Head Island, S.C. 29928.

Bridgeless Daufuskie, just south of Hilton Head, is a stopped-in-time bit of wilderness. Here you can see what the other sea islands were like before causeways opened them to development.

During the American Revolution Daufuskie Island and Hilton Head were, for all intents and purposes, at war. Hilton Head supported the Revolution while Daufuskie islanders were Torries who sided with the British. The two waged bitter battles against each other and it took many years for the hard feelings generated by that struggle to dissipate.

Daufuskie's small population of less than 200 is composed primarily of the descendants of freed slaves. Most made their living oystering until pollution from the Savannah River ruined the oyster beds. With no hope of employment, island children stay only until

Sea Islands of the South

they are out of school, then go to the mainland in search of jobs. The community depends to a large extent on retirement and social security checks to get by. Daufuskie Island and its inhabitants were the subject of Pat Conroy's novel, *The Water is Wide*, and the movie, *Conrack*, based upon that novel.

Daufuskie is not really a tourist destination. In fact, some visitors have reported feeling unwelcome on the island. But it's a scenic spot to cruise by or to visit with an experienced tour guide. The best way to see Daufuskie is by cruise boat from Harbour Town on Hilton Head. While the boat passes the island's shoreline, the captain of the *Vagabond* relates some interesting historic sidelights and provides some insights into the current way of life. (On Hilton Head, call 785-4684 for information about the Vagabond's afternoon or sunset cruises.) A special Daufuskie Island lunch cruise features deviled crab and other low country delicacies for those interested in seeing this unique bit of land. (On Hilton Head, call 785-5236 or 785-5237.)

Public boat service is provided by Beaufort County, primarily for school children and other residents of Daufuskie. The boat, which makes trips between Savannah, Daufuskie and Hilton Head, docks at Hilton Head Harbor near Byrnes Bridge. (For information call Beaufort County Offices at 803-785-6100.)

Shrimper's boat

Islands of Georgia and Florida

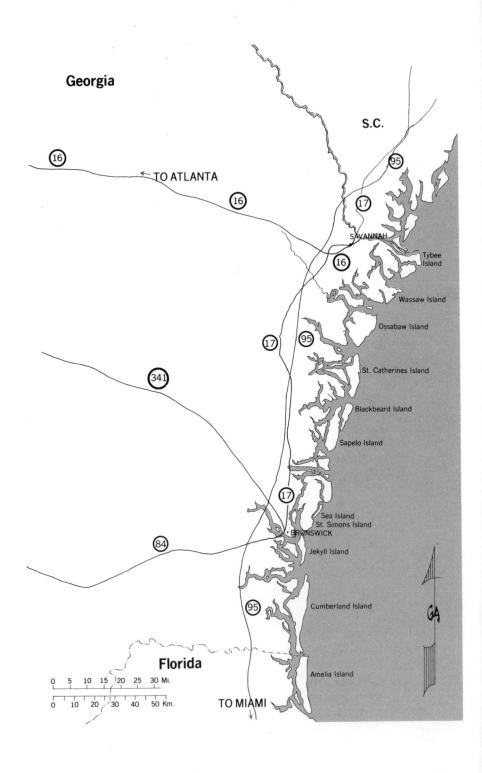

Tybee Island
Savannah Beach

How to get there
Follow U.S. 80 east from Savannah.

Activities and accommodations
Tennis
Swimming: Ocean (showers on boardwalk), hotel/motel pools.
Amusements
Fishing and crabbing
Rental boats
Sightseeing: Historic light, small museum. Fort Pulaski (National Park Service) on Cockspur Island, two miles west of Tybee on U.S. 80.
Accommodations: Motels, hotels, cottage rentals.

For more information
Chamber of Commerce of Savannah, Beach — Tybee Island, P.O. Box 491, Tybee Island, GA 31328.

Tybee (Indian for salt), is 18 miles from Savannah. Georgia's founder, General James Oglethorpe, first touched American soil on this island in 1733. One of the state's oldest resorts, Tybee has long attracted vacationers who come to sun, beachcomb, fish and crab. The boardwalk, fishing pier, amusements and most motels and cottages are clustered near the south end of this four-mile-long island. In the northern section are Tybee Museum, Tybee Lighthouse and the remains of **Fort Screven.** Built in 1875, Screven was manned during the Spanish American War and World Wars I and II. Private homes perch on the walls of this unrestored fort, and **Tybee Museum** is housed in one of its concrete gun placements. The museum offers a folksy rendition of local history from the days of the Indians and Oglethorpe, and on through the Revolutionary and Civil Wars. Highlights include collections of guns and pistols, antique dolls and photographs of life on the island at the turn of the century. (Daily except Tuesday, 1-5, summer hours are 10-6, charge.) **Tybee Lighthouse** completed in 1736 by Oglethorpe, was destroyed by a storm and rebuilt in 1773. For an excellent view of the Savannah River and the Atlantic Ocean, you may climb to the top on weekends and holidays. (2-5, free.)

Sea Islands of the South

Wassaw Island
United States Wildlife Sanctuary

Visitation: None

Wassaw, the most primitive of the Georgia sea islands, was purchased in 1866 for $2500 by George Parson, a New England cotton merchant. His descendants agreed to sell it to the Nature Conservancy provided that no bridge would ever be built connecting the island to the mainland. The Conservancy later deeded Wassaw to the federal government to be used as a wildlife refuge.

This unspoiled gem possesses a variety of natural conditions for the protection and increase of wildlife. Most of the island's ancient dunes are covered with virgin forests of oak, sweet bay, magnolia, cassina holly and slash pine.

Ossabaw Island

How to get there
20 miles south of Savannah. Boat transportation arranged by Turtle Project.
Visitation: Extremely limited and by special permission.

For more information
Turtle Project, 4405 Paulsen St., Savannah, GA 31405.

Ossabaw Island, 20 miles south of Savannah, can only be reached by water. At 43 square miles, it is one of the largest barrier islands. Shell mounds, some with 4000-year-old human bones, are evidence this was once a favorite Indian hunting and fishing ground. The colonists planted indigo on Ossabaw, but the English market for the crop disappeared during the Revolution so they switched to Sea Island cotton. Today all that remains of the plantation era are three former slave dwellings.

In 1961 the island was privately owned by Mr. and Mrs. Clifford West who created the Ossabaw Island Project Foundation for "men and women of creative thought and purpose." The foundation soon gained a reputation as a high quality, if slightly eccentric organization. There's a long waiting list for admission to the select group of scientists, artists, craftsmen, philosophers, photographers, writers and poets who pay approximtely $75 a week to work in this idyllic setting. According to composer Gerald Busby, who completed a new work in three weeks here, "You certainly can get a lot done, and it's extraordinarily beautiful." Others who have taken part have been writer Ralph Ellison and composers Aaron Copland and Samuel Barber. Younger and less established artists live in cabins and grow their own food while enrolled in a related program, Project Genesis.

The foundation considers these creative people endangered like the loggerhead sea turtle which nests on the island. Fearful that development would eventually threaten the foundation as well as the island's wildlife, Eleanor West tried for ten years to protect Ossabaw by selling it to the state. Finally in 1978 the sale was completed for $7 million, a reasonable price considering Arab developers bought smaller Kiawah Island for $17 million two years before.

Some wild residents are definitely not endangered. A herd of Sicilian donkeys has been multiplying ever since Mrs. West gave her son a few of them 14 years ago. They were overgrazing the island to the extent that the state suggested either selling or castrating them. When Mrs. West asked the men on the island which they would prefer if it were them, they voted for vasectomies. A group at the University of Pennsylvania who were interested in studying feral donkeys sent a pair of veterinary surgeons who performed the operations. There have been no complaints. According to Mrs. West, "The donkey wives don't even know."

Ossabaw is bridgeless and destined to remain that way. Most of the purchase price for the island came from a four million dollar gift by philanthropist Robert W. Woodruff with additional donations from Mrs. West and others. Since no federal funds or state recreation money was involved, the danger of uncontrolled public access was eliminated. Under terms of the agreement Mrs. West retained lifetime use of her home and access to certain areas of the island for the Ossabaw Island Project Foundation, which is still supported almost entirely by Mrs. West.

Part of the island is a wildlife refuge which is closed to the public with the exception of those few who participate in a special turtle tagging project. Loggerhead sea turtles return each summer to nest on the island's moonlit beaches. A small research team studies the female turtle's behavior, makes measurements, counts her eggs and attaches an identification tag to her front flipper. If you would like to be a member of such a team, remember you'll have definite responsibilities and be involved in all night beach patrols. But during the day you have the deserted beach to yourself and can explore the maritime forests, creeks, ponds and marshes of this wild and beautiful place.

If this sounds like your type of activity, contact the Savannah Science Museum and sign up for a one week stint as part of a special research team. Lodging is in small cabins and the cost (approximately $180) includes room, board and transportation to the island. The program has only six openings per week. (For infor-

Sea Islands of the South

mation write to Turtle Project, 4405 Paulsen St., Savannah, GA 31405.)

St. Catherines Island

How to get there
South of Savannah via I-95. Boat transportation by individual arrangement.
Visitation: Extremely limited and by invitation only.

St. Catherines Island, 23 wooded square miles, is just south of Ossabaw. The only way to this privately-owned island is by boat and visitation is strictly by invitation. The first coastal Spanish mission was established here in 1566 and an Indian grammar written by a Jesuit friar in 1568 is believed to be the first book written in North America. St. Catherines was the colonial home of Button Gwinnett, a signer of the Declaration of Independence. Forestry and archaeology students from the University of Georgia have done research on this heavily wooded island, which has been designated a National Historic landmark site.

Blackbeard Island
United States Wildlife Sanctuary

Visitation: None.

Blackbeard Island has been the property of the United States government since the 1800s. For a long time it was a quarantine station where ships discharged their passengers who were suspected of having yellow fever. Conditions were less than luxurious if those who described Blackbeard as a hellhole can be believed. The poor souls who didn't have the fever on arrival usually had it before they left thanks to a thriving population of mosquitoes.

The island became a national wildlife refuge in 1924. Today Blackbeard's 5600 acres of woodlands, lakes and marshes are an unspoiled natural haven off limits to the public. Edward Teach (better known as Blackbeard) was supposed to have buried part of his treasure here, but the island's riches are now sea turtles, alligators, quail, turkey, deer and 198 species of birds and migratory fowl that thrive in this lush wilderness.

Georgia & Florida
Sapelo Island

How to get there
The Sapelo Island Ferry leaves from the Meridian, Ga. landing at 9 A.M. and returns at 1 P.M. on Saturdays, leaves 12:30 and returns at 5:30 on Wednesdays. The ride takes 30 minutes. (Charge) The landing is off Georgia 99, eight miles north of Darien on U.S. 17, or ten miles south off the Eulonia exit of I-95.

Visitation: Day visitation only. Extremely limited. Reservations are required.

For more information
For reservations call 912-264-7330. Write: Georgia Bureau of Industry and Trade, P.O. Box 1776, Atlanta, GA 30301.

Sapelo is midway between South Carolina and Florida and may be reached by water from Meridian, a boat landing near Darien. This island, ten miles long and two-to-four miles wide with clean white beaches and large belts of salt marsh, has been open to the public for a short time on a limited basis. If you have advance reservations and the modest fee for a round-trip boat ride (children under six are free), you may join the three-hour tour which leaves at 9 A.M. on Saturday mornings.

After a half-hour ride with sea gulls swooping behind the boat, a bus takes you to the elegant French provincial barn that was once part of the estate of tobacco millionaire R. J. Reynolds. The barn

Mansion on Sapelo Island, Georgia (Photo courtesy of Georgia Coastal Area Planning and Development Commission)

Sea Islands of the South

now serves as main laboratory for the Sapelo Island Research Foundation, which is researching marsh and estuary systems of the coast with the University of Georgia Marine Institute. There you watch a marshlands film which introduces the flora and fauna of the island and also have a chance to examine some real life creatures kept in the lab. Visitors learn how salt marshes form the lifeblood of sea island ecology. Highlights of the guided tour also include a shell-encrusted wild beach, hundreds of sea birds and a massive grove of twisted live-oak trees that were once highly valued for their timber by U.S. Navy shipbuilders. Restrooms and water are available on the island; food and swimming facilities are not.

Sapelo has a fascinating history. The Guale Indians lived here as long ago as 4000 years. Spanish missions were founded in the mid 1500s and freed slaves established Hog Hammock, a permanent settlement which is still there.

The island was purchased by one millionaire after another — Thomas Spalding in 1802, Howard Coffin in 1912 and Richard J. Reynolds in 1933. It was Reynold's love of science which prompted him to establish the Sapelo Island Research Foundation. In '53 he invited the University of Georgia to establish a program of scientific research in marshlands which has attracted scientists from all over the world. In 1976 the Sapelo Island National Estaurine Sanctuary was established by the State of Georgia to preserve the pristine nature of the area. Sapelo is to remain undeveloped and access will be only by boat.

St. Simons Island

How to get there
Car: Turn east off Rte. 17 just north of Brunswick, Ga., onto Rtes. 341/25 to the island.
Boat: Marina facilities.
Plane: Paved airstrip.

Activities and accommodations
Golf
Tennis
Bike Trails
Swimming: Hotel/motel pools, ocean.
Fishing
Horseback riding
Rental boats
Sightseeing: Fort Frederica National Monument, historic sites, small museums.
Accommodations: Hotels and motels.

Georgia & Florida

For more information
St. Simons Chamber of Commerce, Neptune Park, St. Simons Island, Ga. 31522. 912-638-9014.

St. Simons, connected to the mainland at Brunswick by a causeway, is approximately the size of Manhattan. Vestiges of a rich historic background remain on this lovely resort and residential island. In the years since the Indians fished, gathered sea turtle eggs and hunted deer and waterfowl, St. Simons has been under five flags — Spanish, French, British, United States and the Confederate States of America. As on most of the other Georgia Islands, the Spanish built missions and the English built fortifications against the Spanish.

After the Revolution, the success of Sea Island cotton ushered in a new era. This long-fibered cotton which could be easily separated from the seed had been developed in the Bahamas. During the late 1700s and early 1800s, plantations made agricultural history by producing this crop which might bring its owner $100,000 each year.

Planters built fine estates, entertained lavishly and lived a life of luxury. Their opulent lifestyle ended with the Civil War, when Sherman's forces razed the grand plantations. The few people who returned to St. Simons turned to fishing and garden crops, and the island was nearly forgotten until the 20th century. Today St. Simons is being rediscovered by vacationers and new residents whose quest for beauty and ease of pace ends here.

Photo courtesy of St. Simons Chamber of Commerce

Sea Islands of the South

Gascoigne Bluff. Turn left after crossing the causeway from the mainland to explore this wooded, shell-covered bank. King George II originally granted this land to Captain James Gascoigne who, as commander of the *H.M.S. Hawk,* was responsible for bringing the settlers here in 1736. Later it became the estate of one of the island's wealthiest planters. The main shipping wharf for all the island cotton plantations was located here, as was a thriving mill community which cut great live oaks used to build the first U.S. Navy vessels, including the *Constitution* and *Old Ironsides* (1794).

Epworth-by-the-Sea on Gascoigne Bluff, is a conference center for the Methodist Church. Named after Epworth, England, the birthplace of John and Charles Wesley, this lovely site is approached through a grove of moss-draped live oaks and gnarled old cedars. The road winds past two restored slave cabins set in a picturesque 19th century garden. Numerous items relating to the Wesleys and Methodism are displayed in the Methodist Museum.

Retreat Plantation, now the Sea Island Golf Club, is reached by taking King's Way (continuation of the causeway road) and turning right on Retreat Road. The avenue of oaks which leads past the ruins of the slave hospital is a stately reminder of the gracious plantation era. The central part of the golf clubhouse is the old tabby barn; nearby in a grove of trees is the cemetery, where plantation slaves and their descendents have been buried since 1800. Retreat Plantation was once one of the most productive cotton plantations on the coast. When he was a guest at Retreat, Audubon was so impressed he thought he'd landed on "one of the fairy islands said to have existed in the Golden Age."

Slave Hospital at Retreat Plantation (Photo courtesy of St. Simons Chamber of Commerce)

Georgia & Florida

The Lighthouse and Museum of Coastal History are located at 600 Beachview Drive off Ocean Blvd. at the south end of the island. The first lighthouse on this site, erected in 1808, was 75 feet high. It was destroyed by Confederate troops in 1862 to keep its beam from guiding Union invaders to the island. The new light, built in 1872, is one of the oldest navigational beacons still in use on the East Coast. One hundred and six feet above sea level, it has a candlepower of 500,000. The Museum of Coastal History, housed in the restored lighthouse keeper's home, offers exhibits on the economy and lifestyle of the plantation era. Features include an 1870s parlor, folk culture room and crafts cottage. (Tues.-Sat. 9-5, Sun. 1:30-4:30, free.)

At the site of the **Battle of Bloody Marsh,** listen to the self-operated taped message relating the story of this decisive conflict which ended the struggle between Britain and Spain for the southeastern section of the country. In 1742 the Spanish attacked and took possession of Fort St. Simons with a large armada of about fifty ships. 300 conquering Spaniards then set out on foot to take Fort Frederica. Oglethorpe's troops hid in the thick woods and ambushed the unsuspecting Spanish when they stopped to eat. While 200 of the enemy were killed, the British sustained not a single death. Eventually the Spanish sailed back to St. Augustine, having given up the idea of conquering St. Simons. The Battle of Bloody Marsh decided the heritage of future settlers. A different outcome could well have meant you would be reading this book in Spanish.

Christ Church on Frederica Road was first built in 1820. Prior to that time John and Charles Wesley, founders of Methodism in the New World, preached to the Indians and colonists under the trees. The **Wesley Oak** stands near the church. During the Civil War, Union troops were stationed on the island to guard the Inland Waterway. They used the church as a slaughtering place for cattle, broke all the windows and burned the pews for firewood. The ruins stood for more than 20 years until the present structure was built by Anson Green Phelps Dodge, Jr., as a memorial to his wife. The church membership dates to prerevolutionary times, and the cemetery's oldest tombstone bears the date 1803. Visitors are welcome to worship in this picturesque church, which is framed by flowering dogwood and live oaks garlanded with moss. Or they may visit any afternoon. (Winter hours: 1-4, summer 2-5.)

Fort Frederica, at the end of Frederica Road, is one of the largest and most expensive forts ever constructed by the British in North America. Built in 1736 by James Oglethorpe, the fort was used as military headquarters for the defense of Georgia against the Spaniards in Florida. The town of Frederica, planned in England and patterned after an English village, was settled in 1736. The 116 men, women and children who arrived in the New World had been

Sea Islands of the South

carefully selected according to the skills they could perform. The town, protected by earth and timber works with towers and a moat, was destroyed by fire in 1758 and later abandoned. Take time to stroll through the site and examine the tabby ruins of the old fort, remnants of the barracks, tombs in the old burial ground and excavated foundations of some of the original buildings. Markers help visitors visualize the town as it was and give information about its inhabitants. A film about the history of the settlement and the fort is shown at the visitor's center of this national monument. (Open daily from 8-5, free.)

> The world lies east: how ample, the marsh
>   and the sea and the sky!
> A league and a league of marsh-grass, waist-
>   high, broad in the blade,
> Green, and all of a height, and unflecked
>   with a light or a shade,
> Stretch leisurely off, in a pleasant plain,
> To the terminal blue of the main.
>
>   Sidney Lanier

The salt marshes of St. Simons which inspired the poet Sidney Lanier to write the famous *Marshes of Glynn* have a special fascination. Green in summer, gold in autumn and winter, they are a sanctuary for more than 100 species of birds. The marshes,

Fort Frederica (Photo courtesy of St. Simons Chamber of Commerce)

Georgia & Florida

meandering waterways, wide beaches, ancient oaks and year 'round flowers all contribute to the ever-changing beauty of this island.

Sea Island

How to get there
Car: Turn east off Route 17 just north of Brunswick, Ga., onto Rtes. 341/25 to St. Simons Island. Take Sea Island Road over causeway to Sea Island.
Boat: Yacht club facilities available for private boats.
Plane: Paved airstrip nearby.

Activities and accommodations
Golf
Tennis
Biking
Swimming: Pool and ocean.
Fishing: Complete facilities.
Horseback riding
Gun club
Yacht club
Children's activities: Many and varied, well organized.
Sightseeing: Fort Frederica National Monument and historic sites on nearby St. Simons Island.
Accommodations: The Cloister (rooms in hotel, villas, cottages).

For more information
Tourist Division, Georgia Bureau of Industry and Trade, 1400 North Omni International, Atlanta, GA 30303. The Cloister, Sea Island, GA. (800-841-3223 for toll-free reservations).

Sea Island, connected by causeway to St. Simons via Sea Island Road, has had a number of names including Fifth Creek Island, Isle of Palms, Long Island and Glynn Isle. Development of the island began in 1926 when Howard Coffin, an automobile and aviation pioneer, bought five miles of beachfront and decided to establish a first class resort. The opening ceremony in 1928 consisted of invited guests storming the doors for admittance. The host, like the innkeeper in days past, dressed in nightcap, robe and slippers and carrying a lighted candle, welcomed them in and served a sumptuous repast.

Today the elegant **Cloister Hotel** (where gentlemen are still required to wear jackets to all meals including breakfast) sits in carefully landscaped splendor. The resort with its red-tiled roofs

Sea Islands of the South

Marshland and Spanish moss (Photo courtesy of St. Simons Chamber of Commerce)

reflects a strong Mediterranean influence. A dug-out canoe near the entrance serves as a reminder of the once-popular regattas between island plantations.

Guests of the Cloister have rarely complained of nothing to do. The wide range of facilities includes a stable of horses and miles of woodland and beach trails, boats, equipment for salt water fishing, a beach club, fresh-water pools, a gun club, yacht club and dock, tennis, bowling-on-the-green, croquet, a chip-and-putt course, cycling, golf and tennis courts. Five miles of wide private beach beckon the serious sun-worshipper as well as the casual beachcomber. A full program of activities for all ages includes such diverse offerings as art lessons on the beach, disco dancing and sand sculpture.

Jekyll Island

How to get there
Car: Nine miles from Brunswick. Turn east off Route 17 south of Brunswick, Ga. onto Route 50 to Jekyll Island.
Boat: Marina facilities.
Plane: Paved airstrip.

Georgia & Florida

Activities and accommodations
Golf: Four courses.
Tennis
Biking
Swimming: Motel pools, ocean.
Fishing
Boat rentals
Sightseeing: Historic sites including partially restored Millionaires Village.
Camping: Privately operated campground for tents and trailers.
Accommodations: Motels.

For more information
Jekyll Island—State Park Authority, Jekyll Island, GA 31520.

Jekyll Island, eight miles from Brunswick, Georgia, and just south of St. Simons, is the smallest of Georgia's major coastal islands. Ten miles long by one-and-a-half miles at its widest point, Jekyll has 3500 acres of useable land surrounded by 10,000 acres of marsh.

Long before it became *the* social island of the country, Jekyll was a favorite haunt of the Creek Indians who called it Ospo. The French Huguenots in 1562, who named it "Ile de la Somme," were followed by Spanish Jesuit friars who established a mission in 1566. General James Oglethorpe renamed the island for his friend Sir Joseph Jekyll when an outpost was established there under William

St. Simons Island (far left), Jekyll Island (far right), Marshes of Glynn (center), and the town of Brunswick (forefront) (Photo courtesy of Golden Isles Chamber of Commerce)

Sea Islands of the South

Crane Cottage (Photo courtesy of Jekyll Island)

Horton in 1736. Horton's land grant was sold to two other Englishmen before the island was purchased by Christopher du Bignon, a Frenchman who had fought in the American Revolution. For nearly a century, Jekyll remained in the du Bignon family who prospered by using the land to grow Sea Island cotton. In the 1880s a group of northern millionaires chose Jekyll for their winter resort and in 1947, Georgia bought the island for a state park.

The **Millionaires Village** may be reached via Riverview Drive on the west side of the island. The tremendous industrial expansion that took place in the U.S. after the Civil War created many millionaires. Some of the country's wealthiest men commissioned two physicians from Johns Hopkins University to find an ideal winter resort area. Their requirements included a fine climate, pure water, natural beauty and isolation with reasonable proximity to Wall Street. Membership in the famed Jekyll Island Club (formed in 1886) included the Morgans, Pulitzers, Vanderbilts, Rockefellers and Harrimans. For more than 50 years, until 1946, Jekyll Island remained their exclusive playground. During that time "no unwanted foot ever touched the island;" visitors were called "strangers" and were allowed to stay only two weeks. Their "simple life" often consisted of wild-boar hunting parties and ten-course dinners prepared by a chef imported from New York City's Delmonico's Restaurant. Today the village has been restored to

Georgia & Florida

give a glimpse of turn-of-the-century wealth.

A number of buildings in the village are open for visitors; others are undergoing restoration. Bus or walking tours include Moss Cottage (where tickets may be purchased), Goodyear Cottage, Mistletoe Cottage, Rockefeller Cottage, Crane Cottage, Faith Chapel, Jekyll Clubhouse library and Villa Marianna. (Charge)

The **Rockefeller family's 25-room "cottage"** with many of the original furnishings is now the Jekyll Island Museum. This retreat had such extras as towel warmers, a walk-in safe and a Tiffany stained-glass window. (Mon.-Sat., 9-5, Sun., noon-5.)

The **Jekyll Island Clubhouse** is a palatial building which served as the winter residence for the majority of members. Here the 100 families and their guests gathered to dine, play poker and exchange hunting stories. And here it's said J. P. Morgan and his associates founded the Federal Reserve Bank in 1913. In later years a few members built cottages but continued to use the clubhouse for dining and social activities.

Crane Cottage (1916) is the most imposing of the millionaire's homes. In fact the plumbing manufacturer who build the Mediterranean-style structure was criticized for its ostentatiousness. **Mistletoe Cottage** has been restored and, in addition to period fur-

Tiffany stained glass window in Faith Chapel (Photo courtesy of Jekyll Island)

nishings, houses a collection of momentos and photographs from the club's heyday.

Faith Chapel (1904), one of the original buildings in the Village, is a small interdenominational church of English design. Be sure to stop in to view two famous glass windows. One was created by Maitland Armstrong and the other is an Old Testament scene by Louis Comfort Tiffany, said to be one of five he personally installed. (8-dark, daily)

Finding the tranquil beauty of woods, marsh and beach a welcome contrast to the hectic world of business and finance, these magnates often unwound by hunting. Deer, turkeys and quail in Jekyll's forests were supplemented with English pheasant and wild boars which were a gift from the King of Italy. Cold spells, then as now, were brief in the semi-tropical climate, and the club's season ran from New Year's to Easter.

An estimated one-sixth of the world's wealth was controlled from the island during the millionaires' reign. Early in World War II, there was White House concern that too much wealth was concentrated in one place. German submarines had been sighted nearby when the government ordered Jekyll evacuated and placed under protection of the coast guard. This hastened the demise of the Jekyll Island Club. But the younger generation had already focused on other projects and other places and was turning away from one of the most widely known privately owned islands in the world.

As you drive down Riverview Drive, you will pass the ruins of **Horton House,** the former home of William Horton, who established an outpost and plantation on the island in 1738. Horton became major of all the British forces at Fort Frederica after Oglethorpe's return to England. This house, Jekyll's first permanent home, was later occupied by the du Bignon family and was part of their plantation for 100 years.

Across the street is a marker indicating the site of Georgia's first brewery, where beer was made for Oglethorpe's soldiers. The iron kettle on Riverview Drive symbolizes a sad chapter in Jekyll's past. It was retrieved from the last slave ship to land on American soil. The ship, *The Wanderer*, secretly landed a cargo of slaves on the island in 1858 making worldwide news.

When Jekyll was purchased by Georgia, a causeway was constructed which opened the island to development. Today it is maintained as a year-round public vacation facility. Ten miles of uncrowded beaches, four golf courses, tennis courts, bicycle paths, boating, rentals for deep-sea fishing and a campground for tents and trailers have made this a popular place to vacation.

Fortunately you don't have to be a millionaire to take pleasure in the island's natural riches. The salt marshes of cordgrass, needle rush and panic grass are edged by colorful wildflowers—sea oxeye, sea

Georgia & Florida

Cherokee Campground (Photo courtesy of Jekyll Island)

lavender, seaside goldenrod and marsh elder. Forests of magnificent live oaks, magnolia, longleaf and loblolly pines and southern white cedar shelter deer, possum, wild turkey and a wide variety of birds. For those intent on accumulating wealth, the beaches are a favorite hunting ground for sand dollars.

Cumberland Island

How to get there
By National Park Service passenger ferry from St. Marys, Ga., daily, year 'round except Tuesday and Wednesday, Labor Day to Memorial Day. To reach St. Marys, take Ga. 40 east from I-95 near Kingsland. Mainland departure times are 9:15 A.M. and 1:45 P.M. Island departure times are 12:15 and 4:45 P.M. The trip takes 45 minutes. (Charge.)

Activities and accommodations
Shelling
Swimming
Sightseeing: Dungeness ruins.
Camping: One 16-site developed campground and three primitive back-country sites.
Accommodations: Cumberland Island National Seashore. Day visitation only (except for backpackers).

For more information
Ferry reservations are advisable and can be made by calling 912-882-4335 or by writing to the superintendent at P.O. Box 806, St. Mary's, GA 31558.

Sea Islands of the South

Cumberland, Georgia's southernmost coastal island, extends almost to the Florida border. Sixteen miles long and 1½-3 miles wide, the Cumberland Island National Seashore is one of America's last unspoiled beaches. After a 45-minute boat trip, you will probably want to take the short walk to the ruins of **Dungeness Mansion** before going to the beach. Backpackers may select one of 16 developed campgrounds offering restrooms, showers and drinking water or one of three primitive back-country sites. Stays are limited to seven days for campers.

Cumberland represents wilderness at its best. Twenty miles of wide sandy beach freckled with shells, acres of salt marsh and a forested interior dotted with freshwater marshes and sloughs await the outdoor lover.

Although the island is relatively undisturbed, man has left his mark here. Indian shell middens, refuse piles where shells were dumped, have been found that are 4000 years old. Like the other islands, Cumberland was favored for hunting and fishing. The Indians who lived here were the Timucuans, a Florida tribe whose customs and language were different from those of the upper islands. They called the island Missoe, which means sassafras.

Cumberland, like the other sea islands of Georgia, had its Spanish and English periods. The English forts built in 1738 at the ends of the island are completely gone. Smugglers used Cumberland as a hideout. In fact, a small neighboring island is still called Hush-Your-Mouth Island. At one time 1800 lumbermen came annually to fell huge live oak trees used in shipbuilding. Remnants of old slave cabins and a small Negro settlement at Halfmoon Bluff are evidence of the once-thriving plantation era.

Dungeness Mansion was the plantation home of General Nathaniel Greene of Revolutionary fame. "Light Horse Harry" Lee died here and was buried in the Dungeness cemetery. Years later his body was moved to Lexington, Virginia to lie beside his son, Robert E. Lee. Thomas Carnegie rebuilt Dungeness in the early 1900s so his family could winter here. Years were spent building the huge four-story house which was constructed on an ancient shell mound. The tabby walls were six feet thick at the base, there were 16 fireplaces and the house was surrounded by 12 acres of semitropical gardens. No wonder it gained a reputation as the "most elegant residence on the coast." Carnegie hospitality at Dungeness was legendary, but the luxurious way of life ended when the palatial mansion was burned during the days of Reconstruction. One estate from the era still stands, **Plum Orchard,** which was built for George Carnegie around 1900. This 20-room home with indoor swimming pool and squash court is scheduled for restoration and eventually for tours.

Cumberland Island's real fascination lies in its natural resources. Visitors sometimes glimpse some of its diverse inhabitants — alliga-

Georgia & Florida

tors, marsh rabbit, raccoon, mink, beaver, sea turtle, otter, gray squirrel, armadillo, ghost crab, wild turkey, deer, feral hogs, feral horses (said to have been descended from those of the Spaniards), and diminutive donkeys originally imported from Sicily. A great variety of birds includes everything from the marsh wren to the more spectacular water birds. White ibis, herons, wood stork, black skimmers and pelicans are most commonly seen.

A visit to the island puts civilized concerns into perspective. Here life continues on an elemental level. Twice a day the tide washes the beach clean and deposits nutrients in the marshes. The incredible fertility of the salt marsh, with its self-adjusting processes of growth, decay, production and consumption, is nothing less than miraculous. Cool caverns of live oaks exude timelessness. Yet there are daily surprises. The brown and shriveled resurrection fern springs to green life after the slightest shower.

Change is relentless. Things are born and they die. Waves continue to roll in. The morning mist rises from the marsh whether man is there or not. A single stalk of sea oats can begin to form a dune. Railroad vines and pennywort anchor it in place. A slow build-up of land over a period of years may be washed away in one violent storm. And then the process begins again.

Cumberland Island abounds with beautiful wild animals (Photo courtesy of Steve Price, author of *Wild Places of the South*)

Sea Islands of the South

Amelia Island, Florida

How to get there
Car: 32 miles northeast of Jacksonville, Fla., Amelia Island can be reached via Interstate 95 and U.S. 17 which connects to A1A onto the island.

Plane: Fly into Jacksonville International Airport (rental cars, limo service and taxis available). Private planes use Fernandina Airport on Amelia Island which has four runways.

Activities and accommodations
Tennis
Biking
Swimming
Fishing
Horseback Riding
Boating
Children's Activities: Well supervised, full day including meals (Amelia Island Plantation)
Sightseeing: Historic Fernandina Beach, Florida
Camping
Accommodations: Rental cottages, hotels and motels, inn rooms or rental villas (Amelia Island Plantation)

For more information
Amelia Island Fernandina Beach Chamber of Commerce, 102 Centre St., P.O. Box 472, Fernandina Beach, Florida 32034 (904) 261-3248

Amelia Island, 30 miles north of Jacksonville, Florida, is a tonic of sun, sand and surf for those feeling pushed and pulled by life's hectic speed. Wide beaches encourage a slow pace and a dip in the sea guarantees a fresh perspective. Thirteen and one-half miles long by one-fourth to two miles wide, Amelia is the southernmost in the chain of barrier islands known as the Golden Isles.

This choice bit of real estate, bounded by St. Mary's River estuary, the Amelia River and Nassau Sound, has been under eight flags. The Timucuan Indians' simple life was disrupted first by the French in the 16th century and then by the Spanish. The Union Jack was raised in 1763, but the Tories who converged on the island during the Revolutionary War left when Florida was ceded to Spain. A military adventurer claimed the island for a brief time, but was ousted by a pirate who took over ostensibly for Mexico. After the U.S. government took formal possession, the claim was disputed by the Confederacy which raised its flag here during the Civil War.

Georgia & Florida

There's not much fighting going on today, but you can still have all the action you want. Deep sea fishing is great! Amelia is one of the few places in the world where bass as large as 25 pounds can be caught all year, and angling in freshwater creeks, rivers and lagoons is also rewarding. Besides having outstanding golf and tennis facilities, Amelia is one of the last spots on the Atlantic Seaboard where you can ride horseback on the beach.

Sightseers and photographers will have a field day in the nearby town of **Fernandina Beach** with its Victorian architecture and docks lines with shrimp boats.

The **Florida Marine Welcome Station,** beside the shrimp docks on Atlantic Avenue, is the only structure of its kind in the United States. This station on the Amelia River is Florida's entrance to the Intracoastal Waterway. Here you can watch a nautical parade, everything from shrimp boats to elegant yachts, plying the waterway while you enjoy a free cup of orange juice.

The picturesque seaport community of Fernandina Beach is in the National Register of Historic Places. Be sure to stop at the Chamber of Commerce near the dock for walking or driving tour directions. Centre Street's charm is enhanced by attractively landscaped plazas complete with benches.

Stroll past the old Palace Saloon or better yet stop in to inspect its splendid carved mahogany bar which a reliable source has called "the best bar East of Boise." Be sure to sample a mini-platter of fresh boiled shrimp while enjoying the local color in the oldest saloon in Florida.

An example of Amelia Island's Beautiful Victorian architecture

Sea Islands of the South

Your do-it-yourself tour now winds through the side streets past old steamboat gothic and Queen Anne houses dating back as far as 1857. Dinner time? The restored 1878 Steak House has seafood as well as steaks, and you might want to check out the rumor that theirs is the best gumbo in the southeast. For a dizzying dose of local color, don't miss the Eight Flags Shrimp Festival the first weekend in May. Bring plenty of film. The annual Blessing of the Shrimp Fleet is combined with gala festivities which include everything from skydiving to a mock pirates landing.

Fort Clinch State Park on Atlantic Avenue in Fernandina Beach welcomes visitors to the most northeasternly point in Florida. Its European-style brick masonry is unique in this country, and you'll get a splendid view of the Atlantic Ocean and Georgia's Cumberland Island from the fort's ramparts. The park's recreational facilities include fishing, swimming, skin and scuba diving, boating, camping, picnicking and hiking.

Amelia Island Plantation, a low-keyed resort and residential community, takes up 850 acres of this 11,600-acre island. Tennis enthusiasts have the run of 19 courts at Plantation Racket Club. For those who unwind best on a golf course, the resort offers 27 challenging holes which architect Pete Dye claims are the only true links on the East Coast. Winding between sand dunes, ocean and vast expanses of marsh, they are certainly some of the most scenic.

Those who prefer a natural environment to groomed greens and composition courts should follow the **Sunken Forest Trail.** The dense semitropical growth includes palms, magnolias and live oaks festooned with graceful strands of Spanish moss. About a quarter of this resort community has been set aside as a nature preserve where raccoons, opossums, armadillos and alligators have no fear of a bulldozer-delivered eviction notice.

The plantation's offerings run the gamut from a complete health club with sauna, whirlpool, steam and exercise room to a well-supervised children's program. Yet in many ways the resort has retained the natural blessings that inspired an English sailor 400 years ago who described Amelia Island as "marvelously sweet, with both marsh and meadow ground, and goodly woods among."

GLOSSARY

Dunes. Hills of sand formed by the wind.

Grits. Coarsely ground hominy often served for breakfast in the South.

Gullah. Refers to blacks who settled as slaves on the Sea Islands and the coastal regions of South Carolina, Georgia, and north-eastern Florida, or their English dialect which is said to be a linguistic link between Africa, the Antilles and America.

Estuary. Biologically productive area where river current meets ocean tide.

Headboat. Boats which take a large number of fishermen out to drift and bottom fish.

Hunting Islands. Colonists labeled the barrier islands hunting islands because the Indians used them as such.

Hush puppies. Small, unsweetened, cornmeal cake fried in deep fat. Named because they were thrown to keep the dogs quiet at outdoor fish fries.

Indigo. Plant from which blue dye is made or the name for the dye itself. From the 1740s until the Revolutionary War, indigo sold to English textile industry provided main source of income for coastal planters.

Intracoastal Waterway. Completed in 1940, this 2000-mile dredged channel allows small craft to navigate the East Coast in protected waters.

Loggerhead sea turtle. Endangered sea turtle that nests on the islands and beaches of the East Coast.

Low Country. Marshy coastal areas of the Carolinas.

Middens. Refuse heap. Indian shell mounds are sometimes referred to as middens.

Okra. Vegetable served fried or used in soups, stews and gumbo in the South.

Rookery. Breeding place for birds.

Savanna. Grassland region with scattered trees.

Sea Island cotton. Long-stapled cotton raised originally in the Sea Islands, now chiefly in the West Indies.

Sea oats. Attractive plant protected by law because of its role in the stabilization of beaches and dunes.

She-crab soup. Street vendors in Charleston used to charge extra for she-crabs which were preferred over he-crabs because their eggs added a special flavor to the soup.

Shoals. Shallow areas in the ocean made hazardous by moving sand bars.

Spanish moss. Plant of the pineapple family which grows in long festoons from tree branches. Since it draws all its nourishment from the air, it is not a parasite (as commonly thought) and does not kill trees.

Tabby. Building material composed of oyster shells, lime and sand, mixed with salt water.

Yaupon tea. Tea made from Yaupon holly shrub, especially popular in the Outer Banks of North Carolina.

BIBLIOGRAPHY

North Carolina

From Currituck to Calabash by Pilkey, Neal and Pilkey. (N.C. Science & Technology Research Center, 1978.)

Flaming Ships of Ocracoke by C. Whedbee. (John F. Blair, 1971.)

Ghosts of the Carolinas by Nancy Roberts. (McNally & Loftin, 1962.)

Graveyard of the Atlantic by David Stick. (The UNC Press, 1952.)

Legends of the Outer Banks by C. Whedbee. (John F. Blair, 1976.)

The Living Land: An Outdoor Guide to North Carolina by Marguerite Schumann. (Dale Press/The East Woods Press, 1977.)

Outer Banks of North Carolina by David Stick. (The UNC Press, 1958.)

Sea Shells Common to North Carolina by Hugh Porter and Jim Tyler. (UNC Institute of Marine Sciences, Morehead City, 1976.)

South Carolina

Diary of a Kiawah Pioneer by Blanche Brumley. (Kiawah Island Company.)

"Edisto Island" by Nell S. Graydon. (R.L. Bryan Company, Columbia.)

A History of Kiawah Island by John G. Leland. (Kiawah Island Company, 1977.)

Pawley's Island by Prevost and Wilder. (State Printing Company, 1972.)

Georgia

"A Beachcomber's Guide to the Golden Isles" by Bertrand H. Dunegan. (Benedictine Military School, Savannah, 1975.)

Georgia's Land of the Golden Isles by Burnette Vanstory. (University of Georgia Press, Athens, 1970.)

Historic Tybee Island by Margaret Godley. (Tybee Museum Assn.,Savannah, 1976.)

Historic Fiction:

The Beloved Invader by Eugenia Price (J.B. Lippincott, 1965.)

Lighthouse by Eugenia Price. (J. B. Lippencott, 1971.)

New Moon Rising by Eugenia Price. (J. B. Lippincott, 1969.)

General

Birds of North America by Robbins, Bruun and Zim. (Golden Press/Western Publishing Co., 1966.)

Winter Birds of the Carolinas by Michael Godfrey. (John F. Blair, 1977.)

Index

—A—
Alexander Sprunt Memorial
   Sanctuary 111
Amelia Island 148
Amelia Island Plantation 150
American Oyster Catcher 34
Angel Wing 41
Anhinga 35
Ann Parker Tours 121
Artificial fishing reefs 89
Atlantic Beach 74
Auger 41

—B—
Back Bay National Refuge 53
Back Island Safari 107
Bald Head Island 107
Barrier island, definition 17
Battle of Bloody Marsh 137
Baum, Carolista Fletcher 56
Baynard Mausoleum 122
Baynard Ruins 122
Beachwalker Park 104, 108
Beacon 70
Bear Island 14, 76
Beaufort, N.C. 72
Beaufort, S.C. 115
Birds 33-37
Bird Island 88
Blackbeard 68, 132
Blackbeard Island 31, 132
Black skimmer 34
Blockade Runner Museum 34
Blockade Runner Museum 82
Boat service (Daufuskie
   Island) 124
Bodie Island 61
Bodie Island Lighthouse and
   Visitor Center 62
Bodie Island marshes 61
Bogue Banks 74
Brown pelican 111
Brunswick Municipal Airport 47
Buck Hall 99
Bulls Island 97
Burr, Theodosia 56
Buxton Woods Nature Trail 66

—C—
Calabash 89
Callico scallop 43
Calories, seafood 46
Camping, National Seashore 62
Camp See Wee 99
Cape Carteret 76
Cape Fear River 83, 87
Cape Hatteras 66
Cape Hatteras Lighthouse and
   Visitor Center 65
Cape Hatteras National
   Seashore 61
Cape Island 97
Cape Lookout Bight 72
Cape Lookout Lighthouse 73
Cape Lookout National
   Seashore 70
Cape Romain National Wildlife
   Refuge 97
Carolina Beach 82
Carolina Beach State Park 83
Caswell Beach 87
Cedar Island 51, 71
Charleston,
   South Carolina 47, 101, 103
Charleston Gallery 108
Chicamacomico Lifesaving
   Station 65
Christ Church 137
Clam 41
Climate 23
Cloister Hotel 139
Cockle 40
Continental islands 17
Coquina 43
Coquina Beach 61
Core Banks 72
Cormorant 35
Corolla 52
Costeau, Jean-Michel
   Institute 120
Crabbing 45
Crabs 45
Crane Cottage 143
Cruise, Hilton Head 123
Cumberland Island 145
Cumberland Island National
   Seashore 145
Currituck Sound 53
Cypress Swamp Rookery 119

—D—
Dare County Chamber
  of Commerce 53
Daufuskie Island 123
Daufuskie Island lunch tour 124
Deveaux Banks 110
Diamond City 73
Diamond Shoals 66
Duck 52
Dunes, formation of 19
Dungeness Mansion 146

—E—
Emmeline and Hessie,
  restaurant 14
Edisto Beach State Park 112
Edisto Island 111
Egret 37
Elegant disk 41
Elizabeth Gardens 59
Emerald Isle 74
Epworth-by-the-Sea 136
Erosion 18

—F—
Faith Chapel 144
Fernandina Beach 149
Figure Eight Island 80
First Flight Airstrip 54
Florida Air 47
Florida Marine Welcome
  Station 149
Folly Island 103
Fort Clinch State Park 150
Fort Fisher 83
Fort Frederica 137
Fort Macon 74
Fort Mitchel 122
Fort Moultrie 102
Fort Pulaski 129
Fort Raleigh National
  Historic Site 59
Fort Screven 129
Fort Walker 122
Francis Marion National
  Forest 99
Fripp Island 14, 115

—G—
Gascoigne Bluff 136
Gnats 25
Graveyard of the Atlantic 65
Great Heart cockle 40
Green, Paul 60
Gull 34

—H—
Hammock Shop,
  Pawleys Island 14, 97
Hammocks Beach
  State Park 14, 76

Hampton Mariners Museum 72
Hang gliding 13, 57
Harbour Town 124
Harkers Island Ferry 72
Harris, John 57
Hatteras Inlet Ferry 67
Hatteras Island 65
Hatteras Village 67
Heritage Farm 119
Heron 36
Hilton Head 117
Hilton Head Plantation 123
Holden Beach 88
Horseshoe crab 42
Horton House 144
Hunting Island 114
Hunting Island Lighthouse 115
Hurricanes 25
Hurricane precautions 26
Hurst, John 77
Hush-Your-Mouth Island 146

—I—
Ibis 36
Ice House Cafe 108
Insects 24
Island, formation of 17
Island Inn 13
Isle of Palms 99
Isle of Palms Beach and
  Racquet Club 14, 99

—J—
Jasmine Porch 14, 108
Jeep safari 107
Jelly fish sting 24
Jekyll Island 140
Jekyll Island Club 144
Jekyll Island Clubhouse 142
Jockeys Ridge State Park 57
Johns Island Airport 104, 108

—K—
Kiawah Island 104
Kiawah Island Inn 104
Kill Devil Hills 54
Kitty Hawk 53
Kitty Hawk Kites 58
Knobbed whelk 41
Kure Beach 82

—L—
Laura A. Barnes
  shipwreck 62
Lettered olive 41
Lighthouse and Museum of
  Coastal History 137

156

Lighthouses
　Bodie Island 62
　Cape Lookout 73
　Hunting Island 115
　Ocracoke 69
　Old Baldy 85
Lighthouse Island................. 97
Lighthouse Trail................. 115
Loggerhead
　sea turtle 29, 105, 116, 131
Long Beach 87
Lost Colony 59

—M—

Manteo 51, 60
Marsh birds..................... 35
Marsh, ecology of 19
Marsh periwinkle 42
Migration of islands............ 18
Millionaires Village 142
Mistletoe Cottage 143
Morehead City 74
Mosquitoes 25
Moultrie, William 102
Museum of the Sea 66

—N—

Nags Head 52, 55
National Registry of
　National Landmarks 56
New Bern, N.C. 47
North Carolina Marine
　Resources Center
　　Bogue Banks................. 76
　　Kure Beach 83
　　Outer Banks 60
North Pond 64

—O—

Oak Island...................... 87
Oak Island Golf Club 87
Ocean Isle Beach 88
Ocracoke Island 67
　aerial tour and
　　sailing trip 58
　chartered boats 70
　fishing guide service 70
　hunting guide service 70
　inlet 68
　lighthouse 69
　National Park Service
　　Visitor Center 70
　ponies 69
　trolley 69
　village 68
Old Baldy 85
Oregon Inlet 63
Oregon Inlet Bridge 57
Oregon Inlet Fishing
　Center 62

Oregon Inlet Queen 63
Oregon Inlet Restaurant 62
Oristo Resort 113
Osceola 102
Osprey 36
Ossabaw Island................. 130
Ossabaw Island Project
　Foundation 130
Outer Banks 51
　aerial tour 58
　climate 23
Oyster 42

—P—

Pawley, George II 95
Pawleys Island 95
Pawley, Percival................ 96
Pea Island Wildlife Refuge 63
Pelican........................ 34
Pelican Island 70
Pleasure Island 82
Plover 33
Plum Orchard.................. 146
Poe, Edgar Allan 102
Point Harbor 51
Portsmouth Island 58, 70
Portsmouth Village 71

—R—

Raccoon Keys 97
Raleigh, Sir Walter 58
Razor clam 41
Restaurant by George 14
Retreat Plantation 136
Reynolds, R.J. 133
Roanoke Indian Village 60
Roanoke Island 58
Rockefeller's Cottage 143
Rogallo, Francis 58

—S—

St. Catherines Island 132
St. Simons Island 134
Salter Path 76
Sandcastle, (Bodie Island)...... 62
Sand dollar 40
Sandpiper 35
Sapelo Island 133
Sapelo Island Research
　Foundation 134
Savannah....................... 47
Savannah Beach 129
Savannah Science Museum 131
Scotch bonnet 43
Seabrook Island 108
Seabrook Island Resort 109
Sea Island 139
Sea Island cotton 112, 121, 135
Sea Islands, definition of 17
Sea pen 40

157

Sea Pines Forest Preserve 119
Sea Pines Hunter Classic 120
Sea shells 39-43
Sea turtle 29
Shackleford Banks 72
Sharpe, William 77
Shell Castle 70
Shore birds 33
Silver Lake 69
South Brunswick Islands 88
Southport 87
Southport-Fort Fisher Ferry 84
Smith Island 86
Spring Festival 116
Starfish 40
Sullivans Island 101
Sunken Forest Trail 150
Sunray Venus 42
Sunset Beach 88
Swamp Fox nature trails 110
Swan Quarter 51

—T—

Teach, Edward 68, 132
Tern 33
Theodore Roosevelt National
  Area..........................76
Thomas Hariot Nature Trail 59
Tides 26
Topsail Island 78
Transportation hints 47
Travel Venture Tours 121
Turkey wings 42
Turtle Project 131
Tybee Island 129
Tybee Lighthouse 129
Tybee Museum 129

—U—

University of Georgia Marine
  Institute.....................134
U.S. Hang Gliding Association 58
U.S. Lifesaving Service 65

—V—

Vanderhorst Plantation Home 107
Verrazano, Giovanni da 58

—W—

Wassaw Island 130
Water Tour 107
Wesley Oak 137
Whalebone Junction Information
  Center 61
Whooping Crane Pond 119
Wilmington, N.C. 47
Wright, Wilbur and Orville 54
Wright Brothers National
  Memorial.....................55
Wright Brothers National Park
  Service Visitor Center 55
Wrightsville Beach 81

—Y—

Yaupon Beach 87

Other Travel Books from The East Woods Press

Hosteling USA—The Official American Youth Hostels Handbook
Wild Places of the South by Steve Price
Walks in the Great Smokies by Rod & Priscilla Albright
Trout Fishing the Southern Appalachians by J. Wayne Fears
The New England Guest House Book by Corinne Madden Ross
Roxy's Ski Guide to New England by Roxy Rothafel
The Maine Coast—A Nature Lover's Guide by Dorcas S. Miller
Exploring Nova Scotia by Lance Feild
Honky Tonkin'—A Travelguide to American Music
　　　　　　　　　　　　　　　　by Richard Wootton
Steppin' Out—A Guide to Live Music in Manhattan
　　　　　　　　　　　　　　　　by Weil & Singer

Sea Islands
OF THE SOUTH

A Color Portfolio
by Bill Gleasner

10,000-lb. haul of croakers at Corolla

Old Coast Guard station near Corolla on Outer Banks

Hang gliding school in flight, Kill Devil Hill, N.C.

Cape Hatteras Lighthouse

Sea oats at sunrise

A sand street in Ocracoke

Joe Bell flower (Gailardia)

Quaint fishing pier, Silver Lake, Ocracoke, N.C.

View from bridge to Emerald Isle, Bogue Banks, N.C.

Sand patterns on a deserted beach

Sea oats and dunes, Bear Island, N.C.

Lifeguard, Hammocks Beach State Park

Surf casting at sunset, Bear Island

Old Coast Guard house, Bald Head Island, N.C.

Gulls on southern tip (The Point) of Bald Head.

"Old Baldy," lighthouse on Bald Head Island

Interior marsh

Intracoastal Waterway at Southport

Shipbuilding near Holden Beach, S. Brunswick Islands

Shrimpers' net floats

Hard sand beaches of Isle of Palms are ideal for biking

Bike paths wind throughout Isle of Palms Beach & Racquet Club

Beachcombing is a great sport for young and old

First aid for an injured bird, Kiawah Island, S.C.

Builders of sand castles, Kiawah, S.C.

The fleet is in! Mt. Pleasant near Sullivans Island

Vanderhorst Mansion, Kiawah Island

Children's playground at Kiawah

Crabbing is big sport at Kiawah

A breezy day at the beach

Sailing near Beaufort, S.C.

An osprey nesting on bridge near Hilton Head

Harbour Town Lighthouse, Hilton Head, S.C.

Mirage of diamonds from sunlight and sand

View from lighthouse at Harbour Town, Hilton Head

Historic Christ Church, St. Simons Island, Georgia

Entrance to Ft. Frederica, St. Simons Island

Live Oak Avenue on St. Simons Island

The Millionaires Village, Jekyll Island, Georgia

Seagulls sunning on pier, Jekyll Island

Red fox, Cumberland Island, Ga.

Ruins of Carnegie Mansion, Cumberland Island

Feral horses on Cumberland Island beach

A subtropical scene from Amelia Island, Florida

Sunset at Amelia Island

The graceful pose of an egret, Amelia Island